ESSENTIAL RUBBER FORMULARY

ESSENTIAL RUBBER FORMULARY

FORMULAS FOR PRACTITIONERS

V. C. Chandrasekaran

Plastics Design Library

Library of Congress Cataloging-in-Publication Data

Chandrasekaran, V. C.
  Essential rubber formulary : formulas for practitioners / by V.C. Chandrasekaran.
     p. cm.
  ISBN-13: 978-0-8155-1539-5 (978-0-8155)
  ISBN-10: 0-8155-1539-1 (0-8155)
  1. Rubber goods. 2. Rubber chemistry. 3. Chemistry, Technical--Formulae, receipts, prescriptions. I. Title.

  TS1890.C335 2007
  678'.2--dc22

                          2006103071

Printed and bound by CPI Group (UK) Ltd, Croydon, CR0 4YY

Transferred to Digital Print 2011

Published by:
William Andrew Publishing
13 Eaton Avenue
Norwich, NY 13815
1-800-932-7045
www.williamandrew.com

Dedicated to Esther Shine, Jessica, Abigail, and Christa my granddaughters and Arputham my wife

William Andrew Publishing

Sina Ebnesajjad, Editor in Chief (External Scientific Advisor)

Contents

Preface

This book, *Essential Rubber Formulary: Formulas for Practitioners*, has been compiled by a rubber technologist who is well experienced in the rubber industry for over three decades. Most of the formulations given in this book are practical and proven ones based on factory trials, and were used in the manufacture of quality rubber products conforming to relevant specifications. However, minor modifications are necessary, since process conditions such as speed of equipment, temperatures, and production constraints, and cost factors differ from one factory to another. Each formulation is presented with a brief introduction of the products, followed by a short technical note wherever necessary. Alternate formulations are also given for comparative purposes in some cases. These are often used but unpublished formulations, and were selected for their high market potential for small entrepreneurs.

The objective in writing this book is that these workable factory-scale formulations should reach entrepreneurs, rubber chemists, technologists, and students of polymer/rubber technology. The student community, especially, will get an insight into rubber compounding technology and those students will be benefited in designing and developing their own compound formulations for numerous applications of rubber. For the chemists and technologists, this book is useful in that they can start from mid-point instead of from the beginning, which requires quite a bit of elaborate trial and error methods. This book also can be of much help to general readers aspiring to take up the rubber business.

These formulations can be considered as a foundation for chemists to design formulations of their own for hundreds of thousands of rubber products. Wherever trade names are given, as far as possible the chemical names are also given along with them. The trade names referenced in this book are also given in the appendix along with the respective chemical name equivalents. The readers are, however, advised to refer to the technical catalogues of the suppliers of rubber chemicals for their full descriptions.

The formulations are worked out on the basis of parts per hundred rubber (phr) by weight, and in some cases batch weight proportions. The compound formulation of one product can be interchangeable for another product if the duty conditions and physical properties of the products are similar. The mixing cycle is to be strictly followed as per normal practice.

The shelf-life of these compounds before curing should not be more than 60 days at normal temperature and storage conditions.

The author believes that this book can serve as a technical know-how reference for entrepreneurs. While compiling this book, the author presupposed prior knowledge in rubber technology on the part of the readers. The author believes that this book will be a useful guide to research workers, to corporate personnel in the rubber industry, and to rubber traders. Finally, it is hoped that anything about a formula will be intriguing and so a general reader will be curious to read this book.

A few people helped me in this project whom, if I do not thank, I will be failing in my duty. I thank Peter Premkumar and Arun Chellappa, who motivated me to undertake this project a long time ago, and updated my communication infrastructure. I thank Victor Samveda for his timely interventions and useful suggestions. Lastly, I thank Venkatakrishnan Ranganathan who spent several days with me in front of the computer in the preparation of the manuscript.

V. C. Chandrasekaran

PART 1
ABOUT RUBBER

1   Introduction

The term "rubber" is used for any material which when subjected to an external force deforms with a comparatively low load/deflection ratio and regains its original shape quickly and forcibly when the forces applied to it are withdrawn. Based on this definition there are many materials which can be generally classified into the following:

1. Natural rubber
2. Synthetic rubber

Natural rubber is obtained from the bark of the tree *Hevea brasiliensis*, originally discovered in Brazil. The traditional and centuries old method of slitting the bark and letting the milk-like substance drip down as a thick fluid called "latex" still continues to be the sole method of obtaining natural rubber. The history of natural rubber in Brazil is an exciting tale that changed the lifestyle of the world. The Industrial Revolution and the discoveries that followed were reflected in all sectors of human necessities. Automobiles, locomotives, telephones, electricity, and many innovations in engineering and chemical industries changed the topography, customs, and pace of life in towns and cities. Thanks to its multiple uses in the ever-expanding industries, rubber became a commodity that was in demand worldwide.

In 1927 when Reimer and Tiemann published their work on amino acids, it blew open a new avenue for process industries. Rubber technology became an interesting area for study and research work in this field yielded a wide range of new rubbers. Research work on natural rubber yielded polyisoprene which was found to have similar properties as natural rubber. This led to an avalanche of many distinct rubber types based on styrene, isobutylene, butadiene, isoprene, and chloroprene, which in turn heralded the birth of synthetic rubber. The scientific community felt the inappropriateness of the term "rubber" and coined a new term to cover the entire range of man-made rubbers namely "elastomers" or "polymers." A wide variety of synthetic rubbers have been developed, and the production technology for these rubbers was in the hands of global giants such as Dupont, Bayer, Shell, BASF, Goodyear, Polysar, ICI, Dow, and Exxon.

The use of this material became widespread because the characteristics and properties of both natural and man-made rubbers made them useful in almost all sectors namely automobiles, footwear, pharmaceuticals, steel, paper, transport, electrical and electronics, and chemical engineering. Tens of thousands of product types were in use in these sectors. Rubber became a scarce commodity in the United States during World War II. The United States faced an economic crisis due to the shortage in rubber supply. War tanks, warplanes, and warships required huge quantities of rubber for their spare parts. Every factory, home, office, and military facility throughout the United States used rubber. Due to depletion of stocks, the US government banned the use of rubber other than for defense purposes. The post-war situation had similar impacts worldwide. The industrial and manufacturing sectors were forced to develop new methods and innovative compounding technologies for the manufacture of rubber products for critical, noncritical, and consumer needs.

The rubber products manufacturing sector also saw a revolution in compounding and processing technologies for the various types of natural and synthetic rubbers, with the availability of diverse types of rubber ingredients.

It is interesting to know that the first rubber factory in the world was established near Paris in the early 1800s, and the first in England was established by Thomas Hancock during the same period. Charles Goodyear's discovery of vulcanization and Hancock's discovery of mastication revolutionalized the rubber manufacturing industry from its infancy.

The wide range of rubber products manufactured by industries world-wide catered to defense, civil, aviation, railways, agriculture, transport, textiles, steel, health, sports, and practically every conceivable field. Therefore, designing and developing rubber formulations was a tricky and challenging job for the rubber technologist.

2 Brief Notes on Compounding Ingredients

In order to produce a useful end product, the base rubbers are admixed with suitable ingredients. This process is called compounding. The ingredients used for compounding are classified into accelerators, activators, antioxidants, coloring agents, fillers and reinforcing agents, retarders, rubber process oils, softeners, and vulcanizing agents. Certain other classes of products which do not come under the above classes are dusting and anti-tack agents such as talc powder and mold lubricants such as silicones, reclaimed rubbers, and solvents. Addressing a complete treatise on rubber compounding ingredients would be a voluminous task and it is also not within the scope of this book; however, brief notes are provided for readers to acquire a basic idea of each class of ingredients.

The unit "phr" that appears frequently in the formulations listed throughout this book stands for "parts per hundred rubber," which means the parts by weight of an ingredient per hundred parts by weight of rubber. For example, 40 phr of carbon black means 40 parts by weight of carbon black per hundred parts by weight of rubber, where the unit of weight can be either kilograms or pounds.

2.1 Accelerators

Accelerators of vulcanization are classified into organic and inorganic types. Organic accelerators are known to the rubber industry for over a century. Their use in rubber compounding has become universal. Some examples of organic accelerators are hexamine, mercapto-N-cyclohexyl benzothiazole sulfenamide, sodium diethyl dithiocarbamate, tetramethyl-thiuram disulfide, tetramethylthiuram monosulfide, etc. These compounds represent almost the entire range of organic accelerators from moderate to ultra accelerators. Inorganic accelerators such as lime and litharge are used in slow curing products like rubber lining. Accelerators reduce the time required for vulcanization. The benefits of using accelerators are economy of heat, greater uniformity of finished goods, improved physical properties, improved appearance, and better resistance to deterioration.

2.2   Vulcanizing Agents

Sulfur is the most well-known vulcanizing agent. It is easily available in powder and prilled form packed in polyethylene bags. Sulfur vastly improves the properties of raw rubber which is sticky and soluble in solvents. With the addition of sulfur, rubber is converted into a nontacky, tough, and elastic product.

2.3   Activators

Activators help accelerators in the vulcanization process. Zinc oxide and zinc stearates are the most popular activators. Zinc oxide is also a reinforcing filler.

2.4   Antioxidants

Rubber is degraded by oxidation. In order to prevent this, inhibitors are used during rubber compounding. These inhibitors are called antioxidants. The commercially available antioxidants are grouped into amine types and phenolic types. Products derived from amines, mostly aniline or diphenyl-amine, are called staining antioxidants because they tend to discolor non-black vulcanizates on exposure to light, and products derived from phenol are referred to as nonstaining antioxidants.

2.5   Fillers and Reinforcing Agents

Fillers are classified into reinforcing and inert fillers. They can be either black or nonblack fillers. Those which have a pronounced effect on the physical properties of rubbers, such as tensile strength, abrasion resistance, tear resistance, and fatigue resistance, are called reinforcing fillers. Examples are carbon blacks, zinc oxide, magnesium carbonate, china clay, etc. Fillers that do not have an influence on these properties are called inert fillers (e.g., ebonite dust, graphite powder). Nevertheless, they perform a number of useful functions such as increasing chemical resistance, heat resistance, and ease of processing, providing rigidity or hardness to products, and most important of all reducing the cost of the compound. Examples of inert fillers are whiting, barytes, lithopone, talc, etc., apart from ebonite dust and graphite powder.

2.6 Retarders

Retarders are used to prevent premature vulcanization, called scorching of compounds, during processing and storing. During mixing and further processing in a calender, extruder, or molding press, the rubber compound is continuously subjected to heat which results in premature curing or pre-curing. To prevent this retarders are admixed with the compound. Salicylic acid is a well-known retarder for natural rubber compounds, but it activates curing of neoprene compounds. Excessive use of retarders results in porosity in compounds. Many commercial grades of retarders are available.

2.7 Process Oils/Softeners

These materials are added to rubbers primarily to aid in the processing operations such as mixing, calendering, and extruding. They are used along with fillers to reduce the cost of the compound. Peptizing agents are also softeners which increase the mastication efficiency and reduce the Mooney viscosity level to the desired processability. Paraffin waxes with a melting point of approximately 55°C are used as plasticizers. They bloom to the surface and protect ozone sensitive rubbers against cracking under static stress. Various kinds of resins are used as plasticizers; for example, coumarone resins, petroleum resins, high styrene resins, and phenolic resins to name a few. They give excellent flow characteristics to rubber compounds during calendering, extruding, and molding.

3   Some Hints on Rubber Compounding Techniques

Neither natural rubbers nor synthetic rubbers can be used as such. They must be mixed with other chemicals to obtain a balance of properties to suit the end use. This process is called compounding. The chemicals are mixed in specific proportions. In order to arrive at the exact proportions at the parts per hundred rubber (phr) level, various aspects of the end product should be taken into consideration such as working temperature, oxidation, abrasion resistance, chemical resistance, physical properties, and other properties such as oil, fuel, heat, and ozone resistance. The operational parameters such as speed, temperature, and pressure during mixing, extrusion, calendaring, and molding should be studied.

The quality and durability of modern rubber goods have been made possible only by the practical science of rubber compounding. The rubber technologist is responsible not only for the final quality of the product but also for the adaptability of the rubber compound to manufacturing parameters such as temperature, pressure, and forming. It is easy to make a product prototype having almost any desired properties in the laboratory, but it is not always easy to duplicate this on a large scale, although laboratory data is always valuable for process and product development.

The practice of rubber compounding has undergone mind-boggling changes in the last few centuries, thanks to the diverse compounding ingredients that are springing into the market at a fast pace. Today rubber compounding is considered more as a science and technology using which a skilled technologist can predict the ultimate performance of a rubber product with a fair degree of accuracy.

For proper compounding, a thorough knowledge and understanding of the following points are very vital:

1. The type of rubber basically suited for the service required.
2. The processes and their parameters by which the product will be manufactured (i.e., by mixing, extrusion, calendering, molding, and forming).
3. It is necessary to know the complete physical and chemical properties of the compounding ingredients.
4. A comparative study and evaluation of the gum and filled vulcanizates of the various types of rubbers should be undertaken.
5. Intimate knowledge of the end use and its application.

6. Design aspects such as wall thickness, size, shape of the rubber products, and their mechanical and chemical property requirements.
7. Understanding the differences in the usage of rubber compounds for various applications. For example, compounds used for anti-corrosive tank linings are quite different from compounds designed for other dynamic working conditions and mechanical applications.

The basic design considerations of compound formulation for any product are as follows:

1. Choice of the correct ingredients such as accelerators, fillers, and process aids.
2. Viscosity control of the raw rubber during mastication and mixing.
3. Sticking of compound to mill rollers. Suitable softeners should be used to prevent this.
4. Proper temperature differentials of the mill or calender rolls and extruder during processing and their effect on the compound.
5. Tack improvement by the addition of suitable tackifying agents.
6. Preventing scorching by the addition of retarders.
7. Increase in hardness or modulus by selection and addition of suitable fillers.
8. Increase in elasticity by selection of fillers and their proportions and suitable accelerators and antioxidants.
9. For heat resistance, use of low sulfur doses or sulfurless curing systems can be successfully adopted.
10. All ingredients including mineral fillers are to be of high quality and from branded suppliers.
11. Avoid higher temperatures of curing for thicker articles, especially ebonites. Hot water curing or low temperature step curing is advisable in such cases. The accelerator system should be designed accordingly.
12. Knowledge of processing characteristics: The primary step is mastication before mixing, followed by storing and then further processing in the extruder, calender, etc. The processing characteristics will vary from one compound to another. Premastication of the rubbers is advised for efficient mixing of ingredients and where a low Mooney viscosity is required.
13. Proportions of ingredients: The proportions of ingredients mixed with rubber for various applications differ vastly. The approximate quantity range in a compound formulation with

various ingredients for most rubber products is given below, taking the base rubber as 100 parts by weight.

Ingredient	phr
1. Base rubber or blend of rubbers	100.00
2. Vulcanizing agent (sulfur)	0.5–40
3. Accelerator	0.5–5
4. Activator	1.0–5
5. Antioxidant	0.5–2
6. Reinforcing fillers, carbon blacks, and minerals	25.0–200
7. Processing oils	0.0–25
8. Inert fillers	25.0–200
9. Coloring additives	1.0–5

4 Note on Reclaimed Rubber

Reclaimed rubber is a rubber compounding ingredient. Scrap rubber, unlike scrap steel, undergoes a special process before it can be reused and the rubber that is obtained at the end of this process is known as reclaimed rubber. The rubber reclaiming industry is a specialized industry that resembles its parent industry in development and growth. The use of reclaimed rubber in rubber compounding gives several advantages and therefore it deserves the special consideration of all rubber technologists. The advantages of using reclaimed rubber are:

1. It is economic. Reclaim-added compounds are much cheaper. Even in a high quality compound, reclaimed rubber if used in judicious proportions can give considerable savings.
2. Ease of processibility in the mixing mill, extruder, calender, frictioning, and skim coating operations.
3. Reclaimed rubber is uniform and is compatible with most accelerators, and has a retarding effect on Vulcanisation.
4. Faster curing compounds can be made with reclaimed rubber.

The rubber content in reclaimed rubber may vary between 50% and 60%. This has to be taken into consideration during compounding and substituting for new rubber. When compounding with reclaimed rubber, changes and adjustments in the formula become necessary with respect to zinc oxide, accelerator and filler, and plasticizer levels.

The rubber industry is unique in that its scrap can be obtained as its by-product (i.e., as a recycled product namely reclaimed rubber which can be used by itself). Therefore, the rubber industry and the reclaiming industry are interdependent.

5 Rubber Content in Products

The rubber content in a product is significant in that it mostly justifies the quality and the cost of the product. The user of a rubber product is interested more in the quality and the service performance of a product than its composition. It should be remembered that either high rubber content or incorporation of good quality reinforcing fillers is not necessarily indicative of good performance in service. A clay-loaded low rubber content compound can meet the performance and specification requirements of a product appropriately better than a reinforcing filler-loaded high rubber content product. Given below are examples of rubber goods with different rubber content, which may serve to guide the rubber technologist as to what extent he may load a compound.

Product	Rubber Content
Pure gum stock Friction stock Cushion gum stock for Tires Impact abrasion resistant compounds Y-piece stopper in intravenous tube sets in the pharmaceutical industry Toys	80–90% by weight
Tire treads Footwear Ebonite products Belting compounds Sponge rubber products Rubber rolls Tire tubes, flaps, etc.	50–80% by weight
Water bottles Heels Curing bags Soles and soling compounds Abrasion resistant sheets Mechanical and molded rubber products Tank lining	30–50% by weight

Product	Rubber Content
Rubber mats	10–30% by weight
Battery boxes	
Packing gaskets	
Washers	
Seals, rubber rings, etc.	

It can be noted that there is a wide range of rubber content in different products. In some products rubber is the most important ingredient whereas in other products the rubber-like quality is not required and it is in fact only a binding agent for fillers. If the rubber content is very low, the compounds cannot be mixed properly in a mixing mill or in a Banbury internal mixer. Higher levels of plasticizers and premastication will be required in this case. Generally, it is impractical to blend compounds containing less than 10–15% of rubber by weight in a mixing mill.

6 Note on Coloring of Rubbers

The coloring of rubber products is done either for decorative purposes or for user requirements. The colors used in the rubber industry are divided into two classes: organic and inorganic pigments. Both are used extensively. Examples of the first class are iron oxides, cadmium sulfides, chromium oxides, antimony sulfides, and titanium oxides. Certain materials like zinc oxide and carbon blacks are used for their pigmenting powers as well as their reinforcing properties. The term organic pigment owes its origin to its use in the dyestuff industry. A large number of colors are sold under their chemical or general descriptive names by their suppliers.

Whiteness in rubber goods may be produced by using zinc oxide, TiO_2, or a combination of TiO_2 with calcium or barium sulfates. Zinc oxide should be free from lead. Longer vulcanization time and ageing in service cause yellowing of white compounds. Adding a blue coloring agent up to 0.1 phr might offset this effect. Pale crepe rubber is most suited for colored goods. Smoked sheets require additional coloring pigments. Low sulfur compounds are well suited for colored articles because blooming does not occur. Antioxidants stain the colored products, so it is better to avoid antioxidants, especially the staining types. Instead of this, age resistance is achievable with a low sulfur curing system. For colored articles, dark plasticizers such as pine tar, coal tar, and asphalt should be avoided. Instead, light colored plasticizers such as petroleum, light mineral oils, palm oils, and other vegetable oils are appropriate. For bright colors other than white, the white pigments are reduced sufficiently enough to form only a background for the color. Too much white pigment causes the color to be pale, too little impairs brilliancy and thickness. The difficulties encountered in manufacturing colored products are blooming and bleeding, which occur if the color is soluble in the rubber. Fading of color takes place in open stream vulcanization or subsequently in sunlight. The type of accelerator used should also be considered in colored products. Alkaline type accelerators produce a dull shade and might cause colors to fade. Generally, safe accelerators in producing colored products are of neutral or acidic and sulfur bearing types such as tetramethylthiuram disulfide (TMT). Organic colors are used in smaller proportions than mineral pigments. Usually 1 phr or less is enough. It may be a good practice to make color master batches to ensure good dispersion.

A large number of colors are sold in the market under their descriptive chemical names and composition. The subject of coloring of rubbers has been one that has sustained interest for many years. However, an exclusive and extensive literature is not available. The readers can refer to literatures and brochures pertaining to coloring agents for the textile and dyestuff industries along with those offered by rubber color suppliers.

7   Typical Rubber Testing Methods

7.1   Prelude

The relationship between the stress applied to rubber and the resultant strain was an inspiring subject in the early studies of rubber, and soon after scientists discovered the unusual elastic behavior of vulcanized rubber. But they were unable to pursue their studies satisfactorily because of the lack of knowledge on rubber and its composition and the nonavailability of suitable devices to test their investigations. In many cases, when investigating the stress–strain relationships, the composition of rubber was not defined. With all this, the study of tension stress was later extended to the study of compression stress too.

Later in the 19th century, a systematic study of the physical properties of rubber, after adding certain compounding ingredients to rubber, was carried out by Pahl and his co-workers. Through their efforts and those of many later investigators, a foundation was laid for a better knowledge of the fundamental elastic properties of different types of vulcanizates. Many research works were focused on measurement of physical properties rather than direct evaluation of the performance characteristics of the rubber products in service. However, this led to the development of techniques for testing and the introduction of new methods and devices, accelerated ageing tests, testing under simulated service conditions, etc., which were hitherto being continuously improved, simultaneously creating the need for testing of rubber before processing, during processing, and after vulcanization.

Manufacturers of rubber products understood that without any set norms for testing procedures, they were bound to lose their ground against the end users who in turn were actively involved in drafting specifications for rubber products, such as hardness, tensile strength, resilience, elongation, modulus, and many functional properties such as oil and fuel resistance, flame resistance, and chemical resistance.

In the rubber industry the quality control aspects broadly covered the following tests:

1. Test on unvulcanized rubber stocks and
2. Tests on vulcanized rubbers

Plasticity, scorch time, cure rate, and plasticity retention index (PRI) tests were generally done on the compounded unvulcanized rubber. They are called tests for processibility.

16

7.2 Tests on Unvulcanized Rubber Stocks

7.2.1 Plasticity

Standard plasticity tests use either the rotation or the compression principle. The rotation-based test consists of determining the torque necessary to rotate a disk in a cylindrical chamber filled with unvulcanized rubber under specified conditions of speed and temperature namely 2 rpm and 100°C, respectively. A number proportional to this torque is taken as an index of the viscosity of the rubber. The most popular testing device is called the Mooney Viscometer discovered by the American physicist Melvin Mooney nearly a century ago.

The compression test called the Williams Plasticity Test conducted in a Wallace Rapid Plastimeter consists of compressing a cylindrical unvulcanized rubber test piece axially between parallel plates under specified conditions and measuring the compressed height.

7.2.2 Scorch Time and Rate of Cure

The Mooney Viscometer runs continuously as the test temperature rises and gives a curve. "Scorch time" is the time required for the reading to rise usually by 3 or 5 units above the minimum while the upward slope indicates the rate of cure. By convention, "cure rate" is expressed as the time required for the reading to rise from 5 or 35 units above the minimum.

7.2.3 Plasticity Retention Index

This test measures the resistance of raw natural rubber to oxidation (ISO 2930, ASTM 3194). It is an ageing test in which a small disk of rubber is heated in air at 140°F and its plasticity number is compared with that of an unaged disk.

7.3 Tests on Vulcanized Rubbers

7.3.1 Hardness

The most widely used device is a shore hardness meter. The hardness is measured by the depth of indentation caused by a rigid ball under a spring load or dead load, the indentation being converted to hardness degrees on a scale ranging from 0 to 100. The reading from a dead load hardness meter is called International Rubber Hardness Degrees (IRHD). The spring-loaded meter gives Shore A values. The hardness scale from 0 to 100 is chosen such that '0' represents a rubber having an elastic modulus of zero and '100' represents a rubber having infinite elastic modulus.

The IRHD test consists of measuring the difference between the depths of penetration of the ball into the rubber relative to a 'foot' resting on the surface of the rubber under an initial load of 3 grams and a final load of 570 grams. The IRHD test is for all practical purposes the same as the Shore A scale. Shore A conforms to the rubbery stage and Shore D conforms to hard rubbers (ASTM 2240, DIN 53505) namely ebonites.

7.3.2   Tensile Test

The tensile test consists of stretching rubber samples at a uniform speed in a tensile tester and recording the values of stress on the samples and the resulting elongation at more or less regular time intervals. The curve drawn with the elongation on the abscissa and the stresses on the ordinate axis is called the tensile curve. The tensile stress is the ratio of the total force acting on the sample to the initial cross section of the sample. The tensile stress at the breaking point of the rubber sample is called tensile strength. It is defined as the force per unit area of the original cross section, which is applied when the specimen is ruptured. The maximum elongation is called "elongation at break" or "ultimate elongation." With lowering of the temperature, the maximum elongation of the sample is reduced. The rate of stretching affects both the value of the tensile strength and that of the elongation. At varying rates of stretching, it was found that the higher the rate of stretching the greater will be the values of tensile strength and elongation. Compounding ingredients such as fillers, sulfur, accelerators, and plasticizers have a great influence on the tensile curves of different rubber compounds. The load per unit area of the original cross section at a given elongation value is called modulus.

A rubber compound in some cases may be characterized by the values of tensile strength and ultimate elongation. It may also be characterized by its modulus, for example, at 100%, 200%, 300%, or 500% elongation. The value of the modulus can be calculated directly from the stress–strain curve. This modulus has nothing to do with what is commonly known as the modulus of elasticity.

7.3.3   Resilience

Rebound resilience is a very basic form of a dynamic test in which the test piece is subjected to one half cycle of deformation only. It is defined as the ratio of energy of an indentor after impact to its energy before impact expressed as a percentage, and hence in the case where the indentor falls under gravity it is equal to the ratio of the rebound height to the drop height. Resilience can also be measured using a falling weight. A well-known instrument is the shore scleroscope which has a hemispherically headed striker. A number of designs of falling ball apparatus have been used (ASTM D2632, DIN 53512).

7.3.4   Specific Gravity

This is measured in relation to that of water assuming the density of water to be 1 kg/liter. The method involves weighing a test piece in air and water (ISO 2781: BS 903 Part AI).

7.3.5   Abrasion Resistance

The volume of rubber abraded from a specified test piece when subjected to abrasive wear under specified conditions is called abrasion loss. The reciprocal of abrasion loss is called abrasion resistance. There are many methods for determining the abrasion resistance of a rubber product. The ratio of the abrasion resistance of the rubber under test to that of a standard rubber expressed as a percentage is called the abrasion resistance index. For intra-laboratory comparison standardized abrasion test methods are available, such as Method A Dupont Machine (for tire treads, soles, and heals), Method B Dupont Machine with constant torque modification, and Method C Akron Machine (for rubbers with hardness ranging between 55°A to 80°A). The Method C Dunlop Machine is suitable for testing tire treads, conveyor belt cover compounds, etc.

7.3.6   Spark Testing

In rubber lining, spark testing is carried out with a high frequency spark tester to check the continuity of the rubber lining. Pinholes in a rubber-lined tank will expose the underlying metal surface to corrosive chemicals handled in the tank. Therefore, the continuity of the rubber lining is checked with a high frequency spark tester with a voltage varying from 6000 to 20,000 V depending on the thickness of rubber and also on the conductive property of the lining compound. While using the spark tester, it is necessary to ensure that the spark does not remain continuously in one spot, to ensure that a burnt hole may not occur through the lining, because of the continuous sparking.

7.3.7   Accelerated Tests

A number of accelerated ageing test methods have been developed to predict, to a satisfactory extent, the service life characteristics of rubbers. The physical properties such as tensile strength, modulus of elongation, and hardness values are measured before and after ageing of the test samples, in an air oven, ozone chamber, and oxygen chamber, under specified conditions and the changes in the values are noted. Although exact correlation with service performance cannot be made from these values, they certainly do give a reasonable comparison between different compounds and perhaps the service life of the product can be predicted with a fair degree of accuracy by extrapolation of the results (ASTM D573, D865, and D572).

7.3.8   Low Temperature Flexibility

The elastomer-elasticity of vulcanized rubber decreases as the temperature falls until it changes to steel-elasticity at the so-called elasticity minimum or at the second-order transition temperature. The second-order transition temperature is the temperature at which the rubber compound becomes stiff when it is cooled. The stiffening takes place gradually until this second-order transition temperature is reached. On further cooling, the rubber becomes brittle and the temperature at which this happens is known as the brittle point or brittleness temperature (ASTM D746).

At the second-order transition temperature, vulcanized rubber ceases to be elastomer-elastic. Above this temperature rubber is flexible. Therefore, the low temperature flexibility is closely connected with the elasticity. It follows that a vulcanized rubber with a high initial elasticity at room temperature will not freeze until a fairly low temperature is reached. It will therefore have relatively good low temperature flexibility. As a vulcanized rubber with a fairly high degree of cure normally has relatively high elasticity, it is also found to have relatively good low temperature flexibility.

Several tests are used to determine the low temperature flexibility of rubber compositions, which can be used to correlate rubber behaviors at temperatures as low as −40°C to −70°C in air craft and space craft applications as well as other application in cold regions. The most widely used test methods can be referred to in ASTM D1043, ASTM D1053, and ASTM D1329.

7.3.9   Chemical Tests

Many chemical tests are conducted on vulcanized rubbers with a view to study the vulcanizate characteristics and, to some extent, the nature of composition of the compounded rubber. Some of the chemical tests are discussed below.

7.3.9.1   Total Sulfur

Total sulfur denotes the total amount of sulfur present in the sample in any form. This is estimated by converting the sulfur to sulfate ion using a zinc oxide–nitric oxide mixture in the presence of a strong oxidizing agent like bromine. The addition of barium chloride precipitates barium sulfate which is filtered, washed, and dried in a muffle furnace. The weight of sulfate precipitated is measured to find out the sulfur present.

7.3.9.2   Ash Content

The rubber sample is extracted with acetone continuously for about 8 hours and the extract is evaporated to dryness in a crucible. The weight difference will give the ash content. For accurate results, correction for sulfur in the ash is to be taken into account by determining the free sulfur and deducting the result from that of the total sulfur. The ash content generally gives an indication of mineral filler content.

7.3.9.3   Acetone Extract

Acetone completely extracts the following common ingredients such as resins, sulfur, mineral oils, waxes, pine tar, some dyestuffs, organic accelerators, antioxidants, and peptizing agents. Acetone incompletely extracts fatty oils, fatty acids, bitumen, asphalts, etc. Extraction in acetone is carried out continuously for not less than 8 hours and not more than 16 hours, during which time the sample is protected from light. The weight of the dried residue of the extract expressed as a percentage of the weight of the test specimen represents the "acetone extract."

7.3.9.4   Tests for Copper and Manganese

These metals have the tendency to accelerate the deterioration of rubber. The upper limit is 50 ppm for copper and 10 ppm for manganese. The copper in rubber is estimated volumetrically as copper sulfate whereas manganese is estimated gravimetrically as manganese dioxide.

"Where a range is given available observations differ. In most cases the differences are thought to be real, arising from differences in rubber specimens, rather than from errors of observation. Where a single value is given, it is either because no other observations are available or because there seems to be no significant disagreement among values within errors of observation. Where no value is given no data has been found."

—Lawrence A. Wood, National Institute of Science
and Technology, Washington, DC

PART 2
FORMULARY

8 Thin Coatings

8.1 Introduction

Thin coatings are chemical resistant protective coatings used in tanks and vessels for handling food products, alcoholic beverages, and chemicals. Film formation is essential to the performance of these coatings. Solvents play an important role in the film forming process. The consistency and film thickness of the coating, as it dries, is a function of the solvent evaporation and solid content. The following formulations are based on Hypalon and neoprene rubbers.

8.2 The Gray Coating of Hypalon

Ingredient	phr	
1. Hypalon 20	100.00	Part A
2. Stearic acid	2.50	
3. Litharge	30.00	
4. Titanium dioxide	75.00	
5. Fine talc	30.00	
6. Semi-reinforcing furnace (SRF) carbon black	0.50	
7. Phenolic resin	2.50	
8. Chlorinated rubber	15.00	Part B
9. Thiuram E accelerator	0.25	Part C
Total	**255.75**	
Xylene	50.00	
Toluene	50.00	
Hypalon 20	100.00	

Note:
Specific gravity = 1.83.

Mix part A (items 1–7) in a mixing mill and dissolve it in the solvent mixture of xylene and toluene by soaking and stirring. Then add parts B and C which are dissolved in the solvent mixture one after another. For thicker films maintain solids at 40% or increase the number of coats. Adjust the solvent level to improve brushability. Before applying, prepare the metal surface by sand blasting. Shot blasting is not preferred. Use more solvent for evaporation makeup.

8.3 The Black Coating of Neoprene

Ingredient	phr
1. Neoprene AC	100.00
2. Coumarone-indene (CI) resin	10.00
3. Fine China clay	10.00
4. Fine thermal black or SRF black	80.00
5. Magnesium oxide	4.00
6. Antioxidant PBNA (phenyl beta-naphthylamine)	2.00
Total	**206.00**

Note:
Specific gravity = 1.05.

Dissolve the mix in the solvent mixture or toluene or xylene in a ratio of 70 : 30. Adjust the solvent level to reach 30–40% solid content, and add more solvent if needed for evaporation makeup and for smooth brushability. A homogeneous solution will help ease of brushability.

8.4 Black Brushing

Ingredient	kg
1. Neoprene KNR	20.00
2. Antioxidant MC	0.40
3. Tetramethylthiuram disulfide (TMT)	0.20
4. Medium thermal (MT) black	40.00
Total	**60.60**
Xylene	39.40
Total	**100.00**

Note:
Specific gravity = 1.20.

Prepare the above formulation in a "Z" blade mixer. The cement can be used for painting tanks that are exposed to corrosion using a brush. Multiple coats (up to three or four) may be required for >100 μ thickness.

8.5   Gray Brushing

Ingredient	kg
1. Neoprene KNR	20.00
2. Antioxidant MC	0.40
3. TMT	0.20
4. MT black	0.40
5. Titanium dioxide	16.00
6. Fine talc	23.60
Total	**60.60**
Xylene	39.40
Total	**100.00**

Note:
Specific gravity = 1.34.

Prepare the above formulation in a "Z" blade mixer. The cement can be used for painting tanks that are exposed to corrosion using a brush. Multiple coats (up to three or four) may be required for $>100\,\mu$ thickness.

9   Oil Seals and "O" Rings

9.1   Introduction

Oil seals and "O" rings are used in all hydraulic systems, pumps, pistons, pipe connections, etc. An oil seal used for sealing a rotating member is called a rotary seal. "O" rings are also used for a similar purpose. Oil seals are used to protect shafts and bearings from ingress of dirt and foreign matter and egress of oil or grease. An oil seal generally consists of an outer circular metal part and an inner flexible member that does the actual sealing and is bonded to the metal part by chemical adhesive agents. The sealing member is made of rubber, either synthetic or natural as the case may be. The sealing lip of the flexible member is prepared by cutting away the flash that forms at the sealing edge during molding. A fine sealing edge provides adequate pressure on the shaft to prevent leakage.

The "O" rings are light and flexible and under compression they deform to follow the component parts to be sealed. "O" rings have a long life. They provide effective sealing under constant or varying pressure, high vacuum, and high or low temperatures. Rubber "O" rings have a circular cross section. In the molding of "O" rings, the flash line at 180° is avoided and the same is preferably provided at 45° on the cross section so that the sealing face is smooth. This is achieved using a proper mold design.

Natural rubber seals and rings are suitable for air, acetylene, cold ammonia gas, alcohols, inorganic salt solutions, sugar liquors, castor oil, ethyl chloride, ethylene glycol, formaldehyde, formic acid, hydrochloric acid, water, sea water, wines, and spirits. Nitrile rubbers are suitable for organic solvents, petrol, aviation turbine fuels, hydraulic oils, transformer coolants, mineral oils, etc. High acrylonitrile and medium acrylonitrile content nitrile rubbers are available which can be employed based on the functional requirements of seals. Blends of synthetic rubbers are also employed in formulating a compound depending on specific performance requirements. Neoprene rubber seals are also suitable for the fuels mentioned above. Nitrile seals are suitable for high aromatic liquids. For extreme temperatures, fluoro-rubbers (e.g., Viton) and silicone rubber seals are suitable. Thiokol (polysulfide) rubbers are suitable for esters, ketones, lacquers, and UV light in the case of static seals.

9.2   Rotary Seal (Neoprene)—85°A

Ingredient	phr
1. Neoprene WX	100.00
2. Nonox B (antioxidant)	2.00
3. Light calcined magnesia (MgO)	4.00
4. Fine thermal (FT) black	90.00
5. High abrasion furnace (HAF) black	65.00
6. Whiting	20.00
7. Tritolyl phosphate (TTP)	10.00
8. Paraffin wax	1.00
9. Stearic Acid	1.00
10. 2-Mercaptoimidazoline (NA22)	6.50
11. Zinc oxide	5.00
12. Phenol-formaldehyde resin	10.00
Total	**314.50**

Note:
Cure for 10 minutes at 145°C.
Tensile strength = 1900 lb/sq. in. (pounds per square inch—psi); Modulus at 200%
elongation = 1530 psi; Ultimate elongation = 280%; Shore hardness = 85–90°A.

For bonding to metal use rubber–metal bonding agents such as
Chemlok or Desmodur-R.

9.3   "O" Ring (Neoprene)—60°A

Ingredient	phr
1. Neoprene WX	100.00
2. Zinc oxide	5.00
3. Light calcined MgO	4.00
4. Medium thermal (MT) black	30.00
5. General purpose furnace (GPF) black	40.00
6. Paraffin wax	1.00
7. Stearic acid	1.00
8. Nonox B (antioxidant)	4.00
9. Naphthenic oil	4.00
10. 2-Mercaptoimidazoline (NA22)	1.00
Total	**190.00**

Note:
Cure for 10 minutes at 145°C.
Tensile strength = 2200 psi; Modulus at 200% elongation = 1500 psi;
Ultimate elongation = 320%; Shore hardness = 60–65°A.

9.4   Rotary Seal (Nitrile)—60°A

Ingredient	phr
1. Medium–low acrylonitrile rubber (Paracril B)	100.00
2. Sulfur	1.50
3. Zinc oxide	5.00
4. Nonox B (antioxidant)	2.00
5. FT carbon black	55.00
6. HAF carbon black	28.00
7. Dibutyl phthalate (DBP)	5.00
8. Coumarone-indene resin	5.00
9. Mercraptobenzothiazole	1.50
10. Tetraethylthiuram disulfide (TET)	0.10
Total	**203.10**

Note:
Cure for 10 minutes at 145°C.
Tensile strength = 1850 psi; Modulus at 200% elongation = 850 psi;
Ultimate elongation = 400%; Shore hardness = 60–65°A.

9.5   Rotary Seal (Nitrile)—80°A

Ingredient	phr
1. Medium–low acrylonitrile rubber (Paracril B)	100.00
2. Sulfur	1.50
3. Zinc oxide	5.00
4. Stearic acid	1.00
5. Paraffin wax	1.00
6. Nonox B (antioxidant)	2.00
7. FT carbon black	100.00
8. HAF black	65.00
9. TTP	15.00
10. Accelerator tetramethylthiuram disulfide (TMT)	2.30
Total	**292.80**

Note:
Cure for 10 minutes at 145°C.
Tensile strength = 1760 psi; Modulus at 200% elongation = 1060 psi;
Ultimate elongation = 400%; Shore hardness = 80–85°A.

9.6 Rotary Seal (Nitrile)—75°A

Ingredient	phr
1. High acrylonitrile rubber (Hycar 1001)	100.00
2. Sulfur	1.50
3. Zinc oxide	5.00
4. Phil "A" (FEF)	85.00
5. TTP	10.00
6. Stearic acid	0.50
7. Paraffin wax	1.00
8. Accelerator TMT	1.50
Total	**204.50**

Note:
Cure for 10 minutes at 145°C.
Tensile strength = 1865 psi; Modulus at 200% elongation = 1140 psi;
Ultimate elongation = 455%; Shore hardness = 75–80°A.

9.7 "O" Rings (Nitrile)—65°A

Ingredient	phr
1. Medium–high acrylonitrile (Hycar 1002/1042 can be used in a 60 : 40 ratio)	100.00
2. Zinc oxide	5.00
3. Stearic acid	0.50
4. Nonox B	1.00
5. Sulfur	0.50
6. Fast extrusion furnace (FEF) black "A"	37.50
7. TTP	6.00
8. HBS	3.00
9. TMT	3.00
Total	**156.50**

Note:
Cure for 10 minutes at 145°C.
Tensile strength = 1800 psi; Modulus at 200% elongation = 950 psi;
Ultimate elongation = 450%; Shore hardness = 65–70°A.

9.8 "O" Rings (Nitrile 1)—60°A

Ingredient	phr
1. Hycar 1001	100.00
2. Sulfur	1.50
3. Nonox B (antioxidant)	1.00
4. Zinc oxide	5.00
5. FT black (P33)	55.00
6. HAF black (Phil "O")	28.00
7. DPB	5.00
8. Coumarone-indene resin	5.00
9. MBTS	1.50
10. TET	0.15
Total	**202.15**

Note:
Cure for 10 minutes at 145°C.
Tensile strength = 1900 psi; Modulus at 200% elongation = 1100 psi;
Ultimate elongation = 440%; Shore hardness = 60–65°A.

9.9 "O" Rings (Nitrile 2)—60°A

Ingredient	phr
1. Hycar 1001	100.00
2. Sulfur	1.50
3. Nonox B	1.00
4. Zinc oxide	5.00
5. FT black (P33)	40.00
6. HAF black (Phil "O")	15.00
7. TTP	15.00
8. Pine tar	2.00
9. Coumarone-indene resin	2.00
10. Factice (nitrile grade)	5.00
11. MBTS	1.50
12. TET	0.125
Total	**188.125**

Note:
Cure for 10 minutes at 145°C.
Tensile strength = 1890 psi; Modulus at 200% elongation = 950 psi;
Ultimate elongation = 500%; Shore hardness = 60–65°A.

9.10 "O" Ring Compound (Styrene-Butadiene Rubber, SBR)—55°A

Ingredient	phr
1. SBR (non-oil extended)	100.00
2. Process oil	8.00
3. Stearic acid	2.00
4. Zinc oxide	4.00
5. Santoflex AW	1.50
6. Semi-reinforcing furnace (SRF) black	75.00
7. Paraffin wax	1.00
8. DBG (Diphenyl guanidine) (replace with "F")	0.30
9. CBS	1.20
10. Sulfur	2.00
Total	**195.00**

Note:
Cure for 30 minutes at 140°C.
Tensile strength = 2250 psi; Modulus at 300% elongation = 1750 psi;
Ultimate elongation = 450%; Shore hardness = 55–60°A.

9.11 Rotary Seal (Natural Rubber)—85°A

Ingredient	phr
1. Smoked sheet	100.00
2. Ancoplas ER	2.00
3. Stearic acid	1.00
4. Paraffin wax	2.00
5. Phenyl beta-naphthylamine (PBNA)	1.50
6. Zinc oxide	5.00
7. FT black (P33)	120.00
8. Medium Processing Channel (MPC) black or GPF	50.00
9. MBTS	1.50
10. Sulfur	2.75
Total	**285.75**

Note:
Cure for 30 minutes at 140°C.
Tensile strength = 1800 psi; Modulus at 100% elongation = 750 psi;
Ultimate elongation = 250%; Shore hardness = 85–90°A.

9.12 "O" Rings (Natural Rubber) for Pipe Couplings—60°A

Ingredient	phr
1. Smoked sheet	100.00
2. Ancoplas ER (peptizer—mixer of sulfonated petroleum powder products)	2.00
3. Stearic acid	1.00
4. Zinc oxide	5.00
5. PBNA	1.50
6. P33	50.00
7. GPF	20.00
8. Paraffin wax	1.00
9. Sulfur	2.50
10. MBTS	1.25
11. TMT	0.15
Total	**184.40**

Note:
Cure for 30 minutes at 140°C.
Tensile strength = 2200 psi; Modulus at 200% elongation = 1600 psi;
Ultimate elongation = 500%; Shore hardness = 60–65°A.

9.13 Rotary Seal (SBR)—90°A

Ingredient	phr
1. SBR (non-oil extended grade)	100.00
2. Ancoplas ER	5.00
3. PBNA	1.50
4. Stearic acid	2.00
5. Zinc oxide	5.00
6. P33	120.00
7. GPF	75.00
8. MBTS	1.50
9. TMT	0.15
10. Sulfur	2.00
11. Paraffin wax	2.00
Total	**314.15**

Note:
Cure for 20 minutes at 140°C.
Tensile strength = 1750 psi; Modulus at 200% elongation = 1250 psi;
Ultimate elongation = 330%; Shore hardness = 90°A.

9.14   Rotary Seal (Nitrile)—75°A

Ingredient	phr
1. Polysar Krynac 801 (high acrylonitrile)	100.00
2. DBP	10.00
3. PBNA	1.00
4. Zinc oxide	5.00
5. Stearic acid	1.50
6. P33 black	100.00
7. GPF black	10.00
8. MBTS	1.50
9. TMT	0.15
10. Sulfur	1.50
11. Paraffin wax	2.00
Total	**232.65**

Note:
Cure for 20 minutes at 140°C.
Tensile strength = 1850 psi; Modulus at 200% elongation = 1200 psi;
Ultimate elongation = 450%; Shore hardness = 70–80°A.

9.15   "O" Rings (Nitrile)—60°A

Ingredient	phr
1. Polysar Krynac 800 (medium acrylonitrile)	100.00
2. Diethylhexyl phthalate	15.00
3. Zinc oxide	5.00
4. Stearic acid	1.00
5. Pine tar	5.00
6. P33 black	35.00
7. FEF black	35.00
8. PBNA	2.00
9. Sulfur	1.50
10. MBTS	1.00
11. TMT	0.50
Total	**201.00**

Note:
Cure for 30 minutes at 140°C.
Tensile strength = 1900 psi; Modulus at 200% elongation = 1000 psi;
Ultimate elongation = 400%; Shore hardness = 60–65°A.

9.16   Rotary Seal (Blend of Nitrile/SBR)—75°A

Ingredient	phr
1. Polysar Krynac 800 (medium acrylonitrile)	80.00
2. Polysar Krynac NS (SBR)	20.00
3. DBP	10.00
4. PBNA	1.00
5. Zinc oxide	5.00
6. Stearic acid	1.00
7. P33 black	100.00
8. GPF black	10.00
9. MBTS	1.50
10. TMT	0.15
11. Sulfur	1.75
12. Paraffin wax	2.00
Total	**232.40**

Note:
Cure for 30 minutes at 140°C.
Tensile strength = 1760 psi; Modulus at 200% elongation = 1060 psi;
Ultimate elongation = 400%; Shore hardness = 75–80°A.

SBR swells in oil, compensating for shrinkage of the seal due to leaching of oil at high temperatures.

9.17   Rotary Seal (Neoprene)—85°A

Ingredient	phr
1. Neoperne WRT	100.00
2. Light calcined MgO	4.00
3. PBNA	2.00
4. TTP	10.00
5. Paraffin wax	2.00
6. P33 black	100.00
7. GPF	35.00
8. Zinc oxide	5.00
9. Stearic acid	1.00
10. Ethylene thiourea (Robac 22)	0.50
Total	**259.50**

Note:
Cure for 20 minutes at 140°C.
Tensile strength = 1990 psi; Modulus at 200% elongation = 1530 psi;
Ultimate elongation = 280%; Shore hardness = 80–85°A.

9.18   Rotary Seal (Neoprene)—95°A

Ingredient	phr
1. Neoprene GRT	100.00
2. Light calcined MgO	4.00
3. PBNA	2.00
4. Paraffin wax	2.00
5. SRF black	60.00
6. MT black	100.00
7. Stearic acid	0.50
8. Process oil	10.00
9. Robac 22	0.50
10. Zinc oxide	5.00
Total	**284.00**

Note:
Cure for 30 minutes at 140°C.
Tensile strength = 1920 psi; Ultimate elongation = 150%; Shore hardness = 90–95°A.

Acceleration with thiourea is more effective with W type neoprene.

The G type compounds are cheaper but inferior in abrasion, heat resistance, and low temperature properties.

9.19   "O" Ring (Neoprene)—65°A

Ingredient	phr
1. Neoprene WRT	100.00
2. Di-2-ethylhexyl sebacate	6.00
3. Di-2-ethylhexyl phthalate	6.00
4. Dark factice (Factoprene)	10.00
5. Light calcined MgO	4.00
6. Zinc oxide	5.00
7. Stearic acid	1.50
8. Paraffin wax	1.00
9. P33 black	10.00
10. GPF black	35.00
11. PBNA	2.00
12. Robac 22	0.50
Total	**181.00**

Note:
Cure for 20 minutes at 140°C.
Tensile strength = 2200 psi; Modulus at 200% elongation = 1820 psi;
Ultimate elongation = 320%; Shore hardness = 60–65°A.

9.20   Butyl Rubber Seal—75°A

Ingredient	phr
1. Polysar Butyl 301	100.00
2. Zinc oxide	5.00
3. Stearic acid	2.50
4. FEF black	70.00
5. MT black	70.00
6. Paraffin wax	2.00
7. TMT	1.00
8. MBTS	0.50
9. Sulfur	2.00
Total	**253.00**

Note:
Cure for 60 minutes at 140°C.
Tensile strength = 1100 psi; Ultimate elongation = 220%; Shore hardness = 70–75°A.

Butyl compounds are slow curing. Butyl seals are chemically resistant to acid fumes and gases.

9.21   Bromobutyl Seal—70°A

Ingredient	phr
1. Hycar 2202	100.00
2. GPF black	50.00
3. Zinc oxide	5.00
4. Stearic acid	2.50
5. Paraffin wax	2.00
6. TMT	1.00
7. MBTS	0.50
8. Sulfur	2.00
Total	**163.00**

Note:
Cure for 30 minutes at 140°C.
Tensile strength = 2200 psi; Modulus at 200% elongation = 1050 psi; Ultimate elongation = 560%; Shore hardness = 65–70°A.

Bromobutyl rubber is faster curing than normal butyl rubber and has higher heat resistance.

9.22   "O" Ring Thiokol (Polysulfide Rubber) for Airborne Applications

9.22.1   "O" Ring Thiokol—55°A

Ingredient	phr
1. Thiokol FA	100.00
2. Zinc oxide	10.00
3. SRF black	40.00
4. Stearic acid	0.50
5. Diphenyl guanidine (DPG)	0.10
6. MBTS	0.30
Total	**150.90**

Note:
Cure for 40 minutes at 140°C.
Tensile strength = 1200 psi; Modulus at 300% elongation = 750 psi;
Ultimate elongation = 570%; Shore hardness = 55–60°A.

9.22.2   "O" Ring Thiokol—65°A

Ingredient	phr
1. Thiokol ST	100.00
2. SRF black	60.00
3. Stearic acid	3.00
4. *p*-Quinone dioxime	1.30
5. Zinc oxide	0.50
Total	**164.80**

Note:
Cure for 30 minutes at 140°C.
Tensile strength = 1200 psi; Modulus at 200% elongation = 875 psi;
Ultimate elongation = 280%; Shore hardness = 65–70°A.

Resistance to petroleum solvents, esters, ketones, aromatic fuels, oils, greases, lacquer thinners, ozone, sunlight, and ultraviolet rays makes it possible to use polysulfide rubbers for static seals where no other material serves the purpose.

9.23   Typical Nitrile Sealing Formulations for Airborne Applications

Ingredient	Formula 1 (Hydraulic Fluids) (phr)	Formula 2 (Aviation Gasoline) (phr)
1. High acrylonitrile (nitrile rubber)	100.00	100.00
2. Sulfur insoluble	1.50	1.50
3. Zinc oxide	5.00	5.00
4. Stearic acid	1.00	1.00
5. Paraffin wax	1.00	2.00
6. Nonox B	2.00	2.00
7. TTP	15.00	10.00
8. FT black	100.00	—
9. HAF black	65.00	—
10. Accelerator MS	3.50	3.20
11. FEF black	—	85.00
Total	**294.00**	**209.70**

Note:
Cure for 10 minutes at 160°C.
Tensile strength = 1500 psi (formula 1), 1200 psi (formula 2);
Ultimate elongation = 480%, 500%; Shore hardness = 75°A, 65°A.

9.24   Rotary Seal (Hypalon)

9.24.1   Rotary Seal (Hypalon)—85°A

Ingredient	phr
1. Hypalon 20	100.00
2. Epoxy resin	15.00
3. HAF black	55.00
4. MBTS	0.50
5. DOTG	0.25
6. Tetrone A	1.50
7. Polyethylene	2.00
Total	**174.25**

Note:
Cure for 40 minutes at 140°C.
Tensile strength = 2300 psi; Ultimate elongation = 150%; Shore hardness = 85°A.

9.24.2   Rotary Seal (Hypalon)—70°A

Ingredient	phr
1. Hypalon 40	100.00
2. Light calcined MgO	10.00
3. MT black	55.00
4. Stearic acid	1.00
5. Paraffin wax	2.00
6. Tetrone A	2.00
7. Process oil	10.00
Total	**180.00**

Note:
Cure for 40 minutes at 140°C.
Tensile strength = 2300 psi; Modulus at 200% elongation = 950 psi;
Ultimate elongation = 400%; Shore hardness = 65–70°A.

Hypalon 40 is easier to process than Hypalon 20.

9.25   Rotary Seal (Nitrile/PVC Blend)—80°A

Ingredient	phr
1. Paracril Ozo	100.00
2. Diethylhexyl phthalate	15.00
3. Zinc oxide	5.00
4. Stearic acid	1.00
5. Pine tar	5.00
6. P33 black	35.00
7. FEF black	40.00
8. PBNA	2.00
9. Sulfur	1.50
10. MBTS	1.00
11. TMTS	0.50
Total	**206.00**

Note:
Cure for 30 minutes at 140°C.
Tensile strength = 1900 psi; Ultimate elongation = 300%; Shore hardness = 80°A.

9.26 "O" Ring (Nitrile/PVC Blend)—65°A

Ingredient	phr
1. Butakon AC 5502	100.00
2. Zinc oxide	5.00
3. Stearic acid	1.00
4. GPF black	30.00
5. TTP	30.00
6. PBNA	2.00
7. Sulfur	1.75
8. MBTS	1.00
9. TMT	0.50
Total	**171.25**

Note:
Cure for 30 minutes at 140°C.
Tensile strength = 2400 psi; Modulus at 300% elongation = 1600 psi; Ultimate
elongation = 400%; Shore hardness = 60–65°A.

This seal formulation has the base rubber Butakon AC 5502 which is
a blend of acrylonitrile and polyvinyl chloride and as such it has excellent flame resistance due to the presence of PVC in the base Polymer.

9.27 Rotary Seal with Viton for Airborne Applications

Ingredient	phr
1. Viton A-HV	100.00
2. Magnesium oxide	15.00
3. MT carbon black	60.00
4. Copper inhibitor 65 (containing disalicylal propylenediamine)	2.00
5. Diak No. 1	1.25
Total	**178.25**

Note:
Press cure for 30 minutes at 140°C and then post-cure for 24 hours at 200°C in an oven in
steps starting from 100°C.
Tensile strength = 3000 psi; Ultimate elongation = 120%; Shore hardness = 85–90°A.

The copper inhibitor has a retarding effect at processing temperatures
but it activates curing.

Diak No. 1 is the vulcanizing agent—hexamethylenediamine carbonate.

9.28 Nitrile Rubber Ebonite for Oil Resistant Products

Ingredient	phr
1. Butakon S 8551	100.00
2. Sulfur	8.00
3. DPG	1.00
4. Zinc oxide	5.00
5. Stearic acid	1.00
6. Light calcined MgO	5.00
7. Fortafil A-70	30.00
8. China clay	20.00
9. Titanium dioxide	2.00
10. Fast color	0.50
Total	**172.50**

Note:
Cure for 60 minutes in an autoclave at 140°C.
Shore hardness = 95°A.

10 Beltings—Transmission, Conveyor, and V-Belts

10.1 Introduction

V-belts and transmission belts are used for power transmission in almost every industry. These are composite products with nylon or rayon cord reinforcement or cotton duck to provide high strength, high abrasion resistance, minimum stretchability, and longer life. They are generally made of natural or neoprene rubbers where oil exposure is encountered. Nylon/rayon cords are used for reinforcement. They are frictioned with high tack rubber compounds and used as a reinforcing layer in the building and curing of V-belts. Conveyor belts are used in the fertilizer, steel, cement, mining, and transport industries.

Formulations in natural and synthetic rubber for inner layer, cord friction, and skim compounds and cord dipping solutions are given below. The built-up V-belts in pulley moulds are wrapped with high-tension nylon tapes and autoclave cured, whereas conveyor and transmission belts are press cured.

The group of flat rubber belting includes power transmission, conveyor, and elevator beltings. Essentially, the rubber compounds for these beltings surround a skeleton of textile cord or fabric or both. While these reinforcing members support much of the load on the belting, the rubber compound chiefly acts as a protecting cushioning material which shields the skeletal reinforcement against moisture, shock, and other sources of damage.

Power transmitting belts made of rubber having cross sections resembling the letter V are called V-belts. The strength of V-belts is concentrated in the reinforcing cords located in the upper wide portion. These cords are impregnated with rubber. The narrower bottom part of the V-belt has solid rubber which provides flexibility and increased shock absorption.

10.2   V-Belt Inner Layer (Natural Rubber)

Ingredient	phr
1. Smoked sheet	100.00
2. Reclaim rubber WTR	5.00
3. Fast extrusion furnace (FEF) black	10.00
4. General purpose furnace (GPF) black	10.00
5. Mercaptobenzothiazole (MBT)	1.50
6. Zinc oxide	20.00
7. Phenyl beta-naphthylamine (PBNA)	1.00
8. Nonox B (antioxidant)	1.25
9. Stearic acid	2.00
10. Pine tar	0.50
11. Aromatic oil (Dutrex R)	1.00
12. Flexon 310 (naphthenic oil)	0.50
13. Oleic acid	2.00
14. Sulfur	2.00
Total	**156.75**

Note:

Tensile strength = 2900 psi; Modulus at 300% elongation = 1300 psi; Ultimate elongation = 600%; Shore hardness = 55–60°A; Rebound resilience = 45%.

The inner layer in V-belts is extruded as per the required cross section for use in belt building.

10.3   Cord Friction Compound

Ingredient	phr
1. Smoked sheet	100.00
2. Factice	5.00
3. MBT	1.20
4. Philblack A (FEF)	4.00
5. Zinc oxide	5.00
6. Pine tar	2.00
7. Dutrex R (aromatic oil)	10.00
8. Flexon 310 (naphthenic oil)	2.00
Total	**129.20**

Note:

Tensile strength = 2900 psi; Modulus at 300% elongation = 950 psi; Ultimate elongation = 600%; Shore hardness = 55°A.

Nylon cord fabric is frictioned with the above compound in a calendering machine before use in V-belt building as a reinforcing member. The cord is dipped and dried in a cord dipping solution before being frictioned.

10.4   Latex-Based Solution for Cord Dipping

Ingredient	Production Batch (kg)	Trial Batch (kg)
1. Natural latex	40.00	4.00
2. Styrene-butadiene rubber (SBR)–vinylpyridene latex	14.00	1.50
3. Soft water	700.00	70.00
4. Resorcinol	3.00	0.30
5. Formaldehyde	4.50	0.50
6. Sodium hydroxide	0.60	0.10
Total	**762.10**	**76.40**

The above ingredients are mixed homogeneously in a mixer and used as a dipping solution for cord. The dipped cord fabric is dried/cured in a drying chamber before being frictioned with the friction compound.

10.5   Transmission Belting

Ingredient	Friction Compound (phr)	Skim Coating Compound (phr)
1. RMA 4	100.00	100.00
2. Vulcamel TBN	0.20	0.20
3. Zinc oxide	3.00	3.00
4. Stearic acid	1.50	1.50
5. Activated $CaCO_3$	25.00	40.00
6. China clay	—	20.00
7. Vulcatac CH	2.5	1.00
8. Accinox HFN	1.50	1.50
9. Accicure HBS	0.65	1.00
10. Sulfur	3.75	3.75
Total	**138.10**	**171.95**

Note:
Cure for 20 minutes at 150°C.
Tensile strength = 2200 psi (friction compound), 2100 psi (skim coating compound);
Modulus at 300% elongation = 850 psi, 790 psi; Ultimate elongation = 500%, 450%;
Shore hardness = 55°A, 60°A.

10.6 Conveyor Belt Cover Compound (Natural Rubber)

Ingredient	phr
1. Natural rubber RMA 4	100.00
2. Vulcamel TBN	0.15
3. Zinc oxide	3.00
4. Stearic acid	1.50
5. High abrasion furnace (HAF) black	40.00
6. Aromatic process oil	5.00
7. Paraffin wax	0.75
8. Accinox BL	1.50
9. Accinox HFN	1.50
10. Accicure HBS	0.80
11. Accitand A	0.20
12. Sulfur	2.50
Total	**156.90**

Note:

Cure for 20 minutes at 150°C.

Tensile strength = 2500 psi; Modulus at 300% elongation = 1500 psi; Ultimate elongation = 550%; Shore hardness = 60°A; Rebound resilience = 40%.

10.7   Conveyor Belt Cover Compound (Flame Proof)

Ingredient	phr
1. Neoprene W	40.00
2. Natural rubber RMA 4	60.00
3. Zinc oxide	5.00
4. Stearic acid	1.50
5. Zinc borate	15.00
6. Alumina	30.00
7. Chlorinated paraffin wax	25.00
8. Antimony trioxide	5.00
9. HAF black	40.00
10. Accicure HBS	1.20
11. Accitard A	0.10
12. Accinox BL	1.00
13. Accinox ZC	1.00
14. Sulfur	1.80
Total	**226.60**

Note:
Cure for 20 minutes at 150°C.
Tensile strength = 2000 psi; Modulus at 300% elongation = 1500 psi; Ultimate elongation = 400%; Shore hardness = 65°A; Rebound resilience = 35%.

10.8   Conveyor Belt Cover (Natural Rubber/SBR Blend)

Ingredient	phr
1. SBR 1502	75.00
2. Natural rubber RMA 4	25.00
3. Zinc oxide	5.00
4. Stearic acid	2.00
5. HAF black	40.00
6. Aromatic oil	4.00
7. Accinox BL	1.00
8. Accinox 100	0.75
9. Accinox ZC	1.00
10. Accicure HBS	2.00
11. Accicure TMT	1.00
12. Sulfur	0.50
Total	**157.25**

Note:

Cure for 20 minutes at 150°C. Because of the low sulfur cure system, this compound has high heat resistance.

Tensile strength = 1900 psi; Modulus at 300% elongation = 1200 psi; Ultimate elongation = 400%; Shore hardness = 65°A; Rebound resilience = 35%.

10.9   Oil Resistant Raw Edge V-Belt

Ingredient	Base/Cushion (phr)	Friction (phr)
1. Neoprene GRT	100.00	100.00
2. Chopped nylon Fiber	20.00	—
3. Zinc oxide	5.00	5.00
4. High calcined MgO	4.00	4.00
5. Stearic acid	2.00	1.50
6. Accinox DN flake	2.00	1.50
7. Semi-reinforcing furnace (SRF) black	40.00	—
8. HAF black	—	40.00
9. Elasto 710 oil	5.00	30.00
10. Brown factice	—	4.00
11. Coumarone-indene (CI) resin	—	2.00
12. Ethylene thiourea (ETU)	0.25	0.30
13. Sulfur	1.00	—
Total	**179.25**	**188.30**

Note:

Cure for 20 minutes at 150°C.

Tensile strength = 2200 psi (base/cushion), 2500 psi (friction); Modulus at 300% elongation = 1500 psi, 1800 psi; Ultimate elongation = 400%, 450%; Shore hardness = 60°A, 65°A; Rebound resilience = 45%, 40%.

11 Auto Rubber Components (Molded)

11.1 Introduction

Molded rubber components for the automotive industry have a very high market potential. These components are used in both original equipment manufacture and replacement markets. They are made from natural and synthetic rubbers. They are made strictly as per the Society of Automotive Engineers or other international specifications. Some formulations are given below. These formulations can be used for industrial molded rubber components such as washers, gaskets, bushes, diaphragms, and stoppers. Formulations for car mats, flaps, and tubes are also given. Since auto rubber components are generally required to conform to physical property requirements, these details are also given along with the formulations.

11.2 Shock Absorber—55°A

Ingredient	phr
1. SS RMA 1X	100.00
2. Zinc oxide	5.00
3. HFN antioxidant	1.00
4. Stearic acid	1.25
5. Semi-reinforcing furnace (SRF) black	40.00
6. MBTS	1.00
7. Sulfur	2.50
8. HBS accelerator	0.25
9. Aromatic process oil	1.00
Total	**152.00**

Note:
Cure for 30 minutes at 141°C.
Tensile strength = 2700 psi; Modulus at 100% elongation = 270 psi; Modulus at 300% elongation = 1200 psi; Ultimate elongation = 500–570%; Rebound resilience (DIN standard) = 50%; Shore hardness = 55°A.

11.3 Shock Absorber—65°A

Ingredient	phr
1. SS RMA 1X	100.00
2. Zinc oxide	5.00
3. HFN antioxidant	1.00
4. Stearic acid	1.50
5. SRF black	30.00
6. China clay	75.00
7. Aromatic oil	2.00
8. MBTS	1.00
9. HBS	0.25
10. Sulfur	3.00
Total	**218.75**

Note:
Cure for 30 minutes at 141°C.
Tensile strength = 1750 psi; Modulus at 100% elongation = 411 psi; Modulus at 300% elongation = 1158 psi; Ultimate elongation = 420%; Rebound resilience = 43%; Shore hardness = 65°A.

11.4 Shock Absorber 1—60°A

Ingredient	phr
1. SS RMA 1X	100.00
2. Zinc oxide	5.00
3. HFN	1.00
4. Stearic acid	1.50
5. SRF carbon black	37.50
6. China clay	25.00
7. Dutrex R	1.00
8. HBS accelerator	0.25
9. Sulfur	3.00
Total	**174.25**

Note:
Cure for 30 minutes at 141°C.
Tensile strength = 2370 psi; Modulus at 100% elongation = 386 psi; Modulus at 300% elongation = 251 psi; Ultimate elongation = 480%; Rebound resilience (DIN) = 48%; Shore hardness = 60°A.

11.5   Shock Absorber 2—60°A

Ingredient	phr
1. SS RMA 1X	100.00
2. Zinc oxide	5.00
3. HFN accelerator	1.00
4. MBTS	1.00
5. HBS	0.25
6. SRF black	37.50
7. China clay	25.00
8. Dutrex R	2.50
9. Stearic acid	1.50
10. Sulfur	2.75
Total	**176.50**

Note:

Cure for 30 minutes at 141°C.

Tensile strength = 2475 psi; Modulus at 100% elongation = 237 psi; Modulus at 300% elongation = 840 psi; Ultimate elongation = 580%; Rebound resilience (DIN) = 50%; Shore hardness = 60°A.

11.6   Stabilizer Bar Bush—60°A

Ingredient	phr
1. SS RMA 1X	100.00
2. Zinc oxide	5.00
3. HFN	1.25
4. Stearic acid	2.50
5. Dutrex R	3.00
6. SRF black	30.00
7. China clay	25.00
8. General purpose furnace (GPF) black	20.00
9. Sulfur	3.00
10. MBTS	1.00
11. HBS	0.20
12. Paraffin wax	1.00
Total	**191.95**

Note:

Cure for 45 minutes at 148°C.

Tensile strength = 2500 psi; Modulus at 100% elongation = 327 psi; Modulus at 300% elongation = 1200 psi; Ultimate elongation = 500%; Rebound resilience = 33%; Shore hardness = 60°A.

11.7 Stabilizer Bar Bush—67°A

Ingredient	phr
1. SS RMA 1X	100.00
2. Zinc oxide	5.00
3. HFN	1.25
4. Stearic acid	2.50
5. Dutrex R	3.00
6. SRF black	30.00
7. China clay	25.00
8. GPF black	30.00
9. Sulfur	3.00
10. MBTS	1.00
11. HBS	0.20
12. Paraffin wax	1.00
Total	**201.95**

Note:
Cure for 30 minutes at 150°C.
Tensile strength = 2600 psi; Modulus at 100% elongation = 400 psi; Modulus at 300% elongation = 1450 psi; Ultimate elongation = 486%; Rebound resilience = 30%; Shore hardness = 67°A.

11.8 Adhesive Bonding Agent for Fabric Insertion Sheets

Ingredient	phr
1. SS RMA 1X	100.00
2. Zinc oxide	5.00
3. Nonox BL	1.00
4. HBS	0.50
5. Stearic acid	1.00
6. Dutrex R	3.00
7. SRF black	25.00
8. Sulfur	2.75
Total	**138.25**

Note:
Tensile strength = 2800 psi; Modulus at 100% elongation = 140 psi; Modulus at 300% elongation = 625 psi; Ultimate elongation = 648%; Rebound resilience = 46%; Shore hardness = 45°A.

Dissolve 250 g of the compound in 1 liter of solvent oil/petrol.

11.9   Repair Cement for Automotive Belts

Ingredient	phr
1. SS RMA 1X	100.00
2. HFN	1.50
3. MBTS	1.10
4. Zinc oxide	10.00
5. SRF black	10.00
6. Dutrex R	1.00
7. Stearic acid	0.50
8. Sulfur	2.75
Total	**126.85**

Note:
Tensile strength = 2500 psi; Modulus at 100% elongation = 140 psi; Modulus at 300% elongation = 550 psi; Ultimate elongation = 690%; Rebound resilience = 48%; Shore hardness = 45°A.

The compound is dissolved in solvent oil in a 1 : 4 ratio and the resulting solution is used for repairing belts.

11.10   Metal-Bonded Engine Mountings—45°A

Ingredient	phr
1. SS RMA 1X	100.00
2. Zinc oxide	5.00
3. Stearic acid	2.00
4. Antioxidant phenyl beta-naphthylamine	1.00
5. SRF black	30.00
6. Process oil (Elasto 710)	2.00
7. Sulfur	2.50
8. MBTS	1.00
9. HBS	0.25
Total	**143.75**

Note:
Cure for 15 minutes at 140°C.
Tensile strength = 2900 psi; Modulus at 100% elongation = 150 psi; Modulus at 300% elongation = 650 psi; Ultimate elongation = 600%; Rebound resilience = 45%; Shore hardness = 45°A.

The metal plates should be pickled in dilute hydrochloric or sulfuric acid, cleaned with water, and then dried and applied with a coat of chlorinated rubber solution accelerated with an isocyanate-bonding agent such as Desmodur-R. A 1 mm neoprene layer (60°A) is stuck to this coating, then the above compound is loaded and the assembly is cured in a mold in a hydraulic press. The formula for the neoprene inner layer is given elsewhere in this book.

11.11   Tire Flaps—60°A

Ingredient	phr
1. Natural rubber RMA 1X	60.00
2. Whole tire reclaim	80.00
3. Tread crumb	10.00
4. Zinc oxide	4.00
5. Stearic acid	1.50
6. GPF carbon black	20.00
7. China clay	40.00
8. Aromatic process oil	4.0
9. Accinox HFN	1.00
10. Accicure HBS	0.75
11. Accicure TMT	0.20
12. Sulfur	2.00
Total	**223.45**

Note:
Cure for 10 minutes at 150°C.
Tensile strength = 2000 psi; Modulus at 100% elongation = 220 psi; Modulus at 300% elongation = 750 psi; Ultimate elongation = 450%; Rebound resilience = 30%; Shore hardness = 60°A.

11.12   Window Channel Extrusion for Cars (Natural Rubber)

Ingredient	phr
1. RMA 1X	100.00
2. Zinc oxide	3.00
3. Stearic acid	2.00
4. Fast extrusion furnace (FEF) black	50.00
5. Whiting	70.00
6. Microcrystalline wax	2.00
7. Brown factice	5.00
8. Accinox B or BL	1.00
9. Accinox ZDC	0.50
10. Accicure F	1.00
11. Sulfur	2.50
Total	**237.00**

Note:

Cure for 20 minutes at 141°C.

Tensile strength = 2200 psi; Modulus at 100% elongation = 160 psi; Modulus at 300% elongation = 500 psi; Ultimate elongation = 450%; Rebound resilience = 35%; Shore hardness = 65–70°A.

11.13   Window Channel Extrusion for Cars (Styrene-Butadiene Rubber (SBR))

Ingredient	phr
1. SBR 1502	100.00
2. Process oil	10.00
3. Brown factice	10.00
4. Zinc oxide	3.00
5. Stearic acid	1.00
6. FEF black	30.00
7. Whiting	100.00
8. China clay	50.00
9. Coumarone-indene resin	4.00
10. Accicure MBTS	1.00
11. Accicure ZDC	0.20
12. Accinox B/BL	1.00
13. Accinox ZDC	0.50
14. Sulfur	1.80
15. Microcrystalline wax	2.00
Total	**314.50**

Note:
Cure for 15 minutes at 153°C.
Tensile strength = 1800 psi; Modulus at 100% elongation = 1200 psi; Modulus at 300% elongation = 900 psi; Ultimate elongation = 420%; Rebound resilience = 30%; Shore hardness = 65–70°A.

11.14   Neoprene Dust Covers for the Auto Industry—58°A

Ingredient	phr
1. Neoprene WRT	100.00
2. Zinc oxide	5.00
3. Nonox BL	1.50
4. Nonox HFN	0.50
5. Stearic acid	0.50
6. Phil black A	15.00
7. Light calcined MgO	4.00
8. China clay	50.00
9. Vulcafor MS	0.75
10. Sulfur	1.00
11. Dutex RT	8.00
12. Hexaplas PPL	2.00
Total	**188.25**

Note:

Cure for 20 minutes at 153°C.

Tensile strength = 1850 psi; Modulus at 100% elongation = 200 psi; Modulus at 300% elongation = 850 psi; Ultimate elongation = 500%; Rebound resilience = 35%; Shore hardness = 58°A.

11.15   Automotive Tire Tubes—45°A

Ingredient	Natural Rubber (phr)	Butyl Rubber (phr)
1. RMA 1X	100.00	—
2. Polysar 301	—	100.00
3. Zinc oxide	3.00	5.00
4. Stearic acid	2.00	0.50
5. GPF black	—	65.00
6. FEF black	35.00	—
7. Elasto 710	4.00	—
8. Elasto 541	—	10.00
9. Paraffin wax	1.00	—
10. Accinox BL	2.00	—
11. Accinox TQ	—	0.50
12. Accinox ZDC	—	0.50
13. Accicure HBS	1.20	—
14. Accicure MBT	—	0.50
15. Acciure TMT	—	0.10
16. Accitard-A	0.20	—
17. Sulfur	1.50	2.00
Total	**149.90**	**184.10**

Note:
Cure for 15 minutes at 160°C.
Tensile strength = 2400 psi (natural rubber), 2500 psi (butyl rubber); Modulus at 100%
elongation = 150 psi, 170 psi; Modulus at 300% elongation = 700 psi, 750 psi;
Ultimate elongation = 475%, 450%; Rebound resilience = 45%, 40%;
Shore hardness = 45°A, 45°A.

11.16   Low Cost Butyl Tube

Ingredient	phr
1. Butyl 268 (Exxon)	100.00
2. Zinc oxide	5.00
3. FEF black	30.00
4. SRF black	30.00
5. High paraffinic nonstaining process oil	30.00
6. Mercaptobenzothiazole (MBT)	0.50
7. Tetramethylthiuram disulfide (TMT)	1.00
8. Sulfur	2.00
Total	**198.50**

Note:
Paraffinic oil gives lower extrusion shrinkage.
Cure for 30 minutes at 160°C.
Tensile strength = 1800 psi; Ultimate elongation = 450%; Rebound resilience = 35%;
Shore hardness = 45–50°A.

11.17   Car Mat (Natural Rubber)—70°A

Ingredient	phr
1. RMA 5	25.00
2. Tube reclaim	150.00
3. Sulfur	2.50
4. Accicure F	1.20
5. Accicure TMT	0.20
6. Zinc oxide	2.50
7. Stearic acid	1.00
8. Whiting	200.00
9. Barytes	50.00
10. Titanium dioxide	10.00
11. Process oil	15.00
12. Paraffin wax	1.00
13. Nonox SP	1.00
14. Color as desired	1.00
Total	**460.40**

Note:
Cure for 10 minutes at 153°C.
Tensile strength = 1900 psi; Modulus at 100% elongation = 650 psi; Modulus at 300%
elongation = 1750 psi; Ultimate elongation = 500%; Rebound resilience = 35%;
Shore hardness = 70°A.

11.18   Bicycle Tube

Ingredient	Black (phr)	Red (phr)
1. Natural rubber	100.00	100.00
2. Vulcamel TBN	0.15	0.15
3. Zinc oxide	3.00	3.00
4. Stearic acid	2.00	2.00
5. FEF black	25.00	—
6. Whiting	35.00	—
7. Activated $CaCO_3$	—	40.00
8. China clay	—	25.00
9. Process oil	10.00	10.00
10. Brown factice	3.00	3.00
11. Red oxide	—	4.00
12. Sulfur	1.75	1.75
13. Accicure HBS	1.15	1.15
14. Accicure ZDC or TMT	0.15	0.15
15. Accitard A	0.15	—
16. Accinox HFN	1.50	1.50
Total	**182.85**	**191.70**

Note:

Cure for 5–10 minutes at 160°C.

Tensile strength = 2000 psi (black), 1950 psi (red); Modulus at 100%
elongation = 160 psi, 180 psi; Modulus at 300% elongation = 850 psi, 750 psi;
Ultimate elongation = 400%, 400%; Rebound resilience = 35%, 30%;
Shore hardness = 50°A, 45°A.

11.19   Wind Screen Wiper for Automobiles

Ingredient	phr
1. Natural rubber RMA 5	100.00
2. Stearic acid	1.00
3. Activated $CaCO_3$	25.00
4. Whiting	25.00
5. GPF black	5.00
6. Zinc oxide	5.00
7. FEF black	30.00
8. Accinox BL	1.00
9. Accinox ZDC	0.50
10. Accicure F	1.00
11. Sulfur	2.50
Total	**196.00**

Note:

Cure for 5–10 minutes at 153°C.

Tensile strength = 2300 psi; Modulus at 100% elongation = 175 psi; Modulus at 300% elongation = 790 psi; Ultimate elongation = 500%; Rebound resilience = 35%; Shore hardness = 55°A.

11.20 Nitrile Rubber Gasket Molding for Automobiles

Ingredient	phr
1. Hycar 1001	100.00
2. Zinc oxide	5.00
3. Nonox BL	1.00
4. Stearic acid	1.00
5. GPF carbon black	20.00
6. FEF carbon black	20.00
7. China clay	10.00
8. Durtrex R	15.00
9. Hexaplas PPL	10.00
10. Vulcatac CH	2.00
11. Insoluble sulfur	1.50
12. MBTS	1.50
13. Vulcafor MS	0.40
Total	**187.40**

Note:

Cure for 10 minutes at 150°C.

Tensile strength = 1950 psi; Modulus at 100% elongation = 300 psi; Modulus at 300% elongation = 1000 psi; Ultimate elongation = 420%; Rebound resilience = 30%; Shore hardness = 55°A.

This compound has a high dose of plasticizers for molding a softer product. The tackifier is added to improve the building tack which is poor for nitrile rubbers.

11.21   Metal-Bonded Engine Mounting for Automobiles—50°A

Ingredient	phr
1. SS RMA 1X	100.00
2. Zinc oxide	10.00
3. HFN	1.00
4. Nonox BL	0.50
5. HBS	0.80
6. MBTS	0.15
7. SRF black	40.00
8. Vulcatac CH	1.00
9. Dutrex R	1.00
10. Stearic acid	1.00
11. Sulfur	2.75
Total	**158.20**

Note:
Cure for 20 minutes at 150°C.
Tensile strength = 2750 psi; Modulus at 100% elongation = 200 psi; Modulus at 300% elongation = 750 psi; Ultimate elongation = 600%; Rebound resilience = 45%; Shore hardness = 50°A.

The metal plates should be pickled, cleaned with water, and then dried and applied with a coat of chlorinated rubber solution accelerated with isocyanate. A 1 mm neoprene inner layer is applied to this coating, then the above compound is loaded and the assembly is cured in a mold.

11.22   Head Lamp Gasket for Automobiles (Nonstaining)—55°A

Ingredient	phr
1. RMA 1X	100.00
2. Nonox WSP	1.00
3. Zinc oxide	5.00
4. China clay	37.50
5. Whiting	37.50
6. Sulfur	2.75
7. MBTS	1.00
8. TMT	0.20
9. Stearic acid	2.00
10. Paraffin wax	1.00
11. High abrasion furnace (HAF) black	0.50
Total	**188.45**

Note:
Cure for 10 minutes at 153°C.
Tensile strength = 1850 psi; Modulus at 100% elongation = 150 psi; Modulus at 300% elongation = 850 psi; Ultimate elongation = 400%; Rebound resilience = 35%; Shore hardness = 55°A.

11.23   Basic Formula for General-Purpose, Heat Resistant Gasket (Natural Rubber Without Sulfur)—60°A

Ingredient	phr
1. RMA 4	100.00
2. Stearic acid	1.00
3. Nonstaining antioxidant	3.00
4. Zinc oxide	50.00
5. SRF carbon black	75.00
6. TMT	3.00
Total	**232.00**

Note:
Cure for 10–15 minutes at 153°C.
Tensile strength = 3500 psi; Modulus at 100% elongation = 110 psi; Modulus at 300% elongation = 400 psi; Ultimate elongation = 700%; Rebound resilience = 48%; Shore hardness = 55°A.

This formulation can also be used for cast iron/asbestos pipe couplings.

11.24 Basic Formula for Oil Resistant Gasket from Natural Rubber—65°A

Ingredient	phr
1. RMA 4	100.00
2. Antioxidant	1.00
3. Zinc oxide	5.00
4. Lamp black or SRF black	200.00
5. Sulfur	3.00
6. MBTS	1.50
7. Stearic acid	2.00
8. Process oil	3.00
Total	**315.50**

Note:
Cure for 10–15 minutes at 153°C.
Tensile strength = 3000 psi; Modulus at 100% elongation = 100 psi; Modulus at 300% elongation = 500 psi; Ultimate Elongation = 650%; Rebound resilience = 45%; Shore hardness = 55°A.

11.25 General-Purpose Auto Rubber Bush

Ingredient	phr
1. SS RMA 1X	100.00
2. HFN	1.00
3. Zinc oxide	5.00
4. HBS	0.75
5. Stearic acid	3.00
6. Dutrex R	5.00
7. Sulfur	2.25
8. Channel black or GPF	45.00
Total	**162.00**

Note:
Cure for 60 minutes at 150°C.
Tensile strength = 2750 psi; Modulus at 100% elongation = 345 psi; Modulus at 300% elongation = 1300 psi; Ultimate elongation = 450%; Shore hardness = 64°A; Rebound resilience = 31%.
Replacing channel black with furnace black will give higher modulus, increased hardness and resilience, and higher tensile strength.

12　Retreading Rubber Compounds and Cements

12.1　Introduction

Tire retreading is a prospective industry. Old tires of cars, trucks, and other transport vehicles are repaired and retreaded with suitable tread compounds to give satisfactory mileage. The components that are used in the retreading industry are tire tread compound (camel back, slab), under tread strips, cushion gum, fractioned cord fabric, and vulcanizing cement. Tire tread compounds are mostly made of natural rubber and styrene-butadiene rubbers with good abrasion and fatigue resistance. The other layers above the carcass of a tire are repaired with cushion gum compound and under tread strips. Cushion compound has less fillers; under tread strips are resilient and the tread has high abrasion resistance. Some often-used formulations in the retreading industry are given below. Some laboratory scale formulations with different doses of accelerators, process oils, and filler loadings are also given for comparative study.

12.2　Tire Tread or Camel Back 1

Ingredient	phr
1.　SS RMA 1X	100.00
2.　Nonox HFN	1.50
3.　Zinc oxide	5.00
4.　Vulcafor BSO	0.65
5.　Phil black "O" high abrasion furnace (HAF)	45.00
6.　Dutrex R	6.00
7.　Stearic acid	3.00
8.　Sulfur	2.25
Total	**163.40**

Note:
Cure for 60 minutes at 153°C in retreading molds.
Tensile strength = 2500 psi; Modulus at 100% elongation = 200 psi; Modulus at 300% elongation = 900 psi; Ultimate elongation = 500%; Shore hardness = 59°A; Rebound resilience = 30%.

The compound is mixed and extruded in a suitable extruder with a camel back/slab die profile. The accelerator dose is to be monitored as and when required to prevent scorching during extrusion.

12.3   Tire Tread or Camel Back 2

Ingredient	phr
1. SS RMA 1X	100.00
2. Nonox HFN	1.50
3. Zinc oxide	5.00
4. Vulcafor BSO	0.65
5. Phil black "O" HAF	45.00
6. Dutrex R	5.00
7. Stearic acid	3.00
8. Sulfur	2.25
Total	**162.40**

Note:
Curing time is the same as for formula 1.

A small reduction in process oil in compound formula 1 can give a slight increase in resilience. All other physical data are the same as for formula 1.

12.4   Tire Tread or Camel Back 3

Ingredient	phr
1. SS RMA 1X	100.00
2. Nonox HFN	1.50
3. Zinc oxide	5.00
4. Vulcafor BSO	0.60
5. Phil black "O" HAF	45.00
6. Dutrex R	5.00
7. Stearic acid	3.00
8. Sulfur	2.25
Total	**162.35**

Note:
Curing time is the same as for formula 1.
Shore hardness = 59°A.

A marginal reduction in the accelerator level compared with formulas 1 and 2 increases scorch resistance during extrusion. All other physical data are similar except a marginal increase in modulus.

12.5   Tire Tread or Camel Back 4

Ingredient	phr
1. SS RMA 1X	100.00
2. (Blend of arylamines) Nonox HFN	1.50
3. Zinc oxide	5.00
4. Vulcafor BSO	0.60
5. Phil black "O" HAF	45.00
6. Dutrex R	5.00
7. Sulfur	2.25
8. Salicylic acid	0.75
9. Stearic acid	3.00
Total	**163.10**

Note:
Curing time is the same as for formula 1.
Shore hardness = 59°A.

The inclusion of salicylic acid reduces scorching tendency. However, if the salicylic acid level is high, it can produce porosity in the extruded profile. If the extrusion speed and temperature are high, the level of salicylic acid can be reduced from 0.75 phr to 0.5 phr. The values of physical data are expected to be similar to those for formula 3.

12.6   Tire Tread or Camel Back 5

Ingredient	phr
1. SS RMA 1X	100.00
2. Nonox HFN	1.00
3. Zinc oxide	5.00
4. Vulcafor BSO	0.50
5. Sulfur	2.25
6. Vulcatard A (N-nitrosodiphenylamine)	1.50
7. Dutrex-R	5.00
8. Stearic Acid	3.00
9. Phil black "O" HAF	45.00
Total	**163.25**

Note:

Tensile strength = 2700 psi; Modulus at 100% elongation = 200 psi; Modulus at 300% elongation = 950 psi; Ultimate elongation = 570%; Shore hardness = 55°A; Rebound resilience = 32%.

In this formula, the accelerator level is reduced from 0.6 phr, as in formula 4, to 0.5 phr and a retarder is included, as a result of which scorching is easily reduced. Hardness is reduced to 55°A.

To improve building tack, a tackifying agent can be used.

12.7   Tire Tread or Camel Back 6

Ingredient	phr
1. SS RMA 1X	100.00
2. Vulcatard A	0.50
3. Zinc oxide	5.00
4. Dutrex R	5.00
5. Phil "O" HAF	45.00
6. Vulcafor HBS	0.50
7. Sulfur	2.25
8. Nonox BL	1.00
9. Nonox HFN	0.50
10. Stearic acid	2.50
Total	**162.25**

Note:
Cure for 60 minutes at 150°C.
Tensile strength = 2300 psi; Modulus at 100% elongation = 200 psi; Modulus at 300% elongation = 950 psi; Ultimate elongation = 540%; Shore hardness = 53°A; Rebound resilience = 33%.

With the change of accelerator from BSO to HBS, the following physical properties are obtained. The compound has better scorch resistance.

Resilience can be increased, hardness and modulus can be reduced, and scorching is reduced by keeping the accelerator level low.

In all cases of tread compounds, temperature differentials in the extruder barrel, head, and at the die head are to be maintained suitably as per normal practice during the extrusion operation.

12.8   Under Tread Strips

In any of the above tread formulations, a general purpose furnace black such as Phil black G can be used at 40 phr replacing the HAF black (45 phr) to make a compound for under tread strips. A tackifying agent such as gum rosin can be admixed along with process oil to the compound to improve building tack and processibility in calendering operations.

12.9   Cushion Gum Compound

A cushion gum compound contains semi-reinforcing furnace (SRF) filler as listed in the formulation below:

Ingredient	phr
1. SS RMA 1X	100.00
2. Zinc oxide	5.00
3. Nonox BL	1.00
4. Stearic acid	2.00
5. Dutrex R	3.00
6. SRF black	30.00
7. Sulfur	3.00
8. MBTS	1.00
9. HBS	0.25
Total	**145.25**

Note:
Tensile strength = 3000 psi; Modulus at 100% elongation = 175 psi; Modulus at 300% elongation = 550 psi; Ultimate elongation = 650%; Rebound resilience = 50%; Shore hardness = 40–45°A.

For cushion gum, the raw rubber is to be premasticated well to a low Mooney viscosity (of about 30) to retain building tack.

12.10   Vulcanizing Solution

Ingredient	phr
1. SS RMA 1X	100.00
2. Nonox HFN	1.50
3. MBTS	1.10
4. Zinc oxide	10.00
5. SRF black	10.00
6. Dutrex R	1.00
7. Stearic acid	0.50
8. Sulfur	2.75
Total	**126.85**

This compound being a solution compound has a higher dose of sulfur and low levels of zinc oxide and SRF as filters. This is necessary to keep the rubber content high for bonding and to allow sulfur to bloom out from the film of rubber formed after evaporation of the solvent to other layers such as cushion gum and under tread strips during retreading of rubbers. The use of MBTS improves scorching properties. The mixed compound is dissolved in solvent oil/special boiling point spirit to make a vulcanizing cement for retreading purposes. The normal solid content is about 30%.

13 Industrial Rubber Rollers

13.1 Introduction

Rubber rollers are required for various industries such as textile mills, paper mills, glass plants, steel mills, printing industries, tanneries, mining, and many other industrial applications. The manufacture of rubber rollers involves special skills and know-how. It involves laying of green rubber sheets on metallic cores to the required thickness, and then the composite is wrapped with wet cotton or nylon tapes and vulcanized in an autoclave under the required pressure and temperature. The wet cotton or nylon fabric gives the wrapping pressure on shrinkage. After curing, the rolls are ground to a smooth finish to the required dimensions. Rolls with compounds of various hardness levels are manufactured as required by users. The rapid growth of engineering and chemical industries worldwide has increased the demand for rubber rollers. These rollers are manufactured from either natural or synthetic rubbers. Readers can compare the roll compounds with respect to the role of sulfur, mineral and black fillers, and accelerators (both organic and inorganic types) and their properties. The formulas given below are named with arbitrary notations for identification.

13.2 Method

Rubber stocks are compounded as per the formulations by mill mixing according to the specific need. The metal cores are sand/shot blasted to give a white metal surface. After blasting, a primer rubber coating with a metal rubber bonding agent is applied to preserve the blasted surface and to ensure better rubber to metal bonding. Then the cores are covered with calendered rubber sheets and wrapped with nylon/cotton fabric. The covered rolls are then vulcanized in an autoclave. The vulcanized rolls are then ground and polished. Sometimes an ebonite underlayer is applied on the coated metal for good bonding.

13.3   Cylinder 38—Paper Mills

Ingredient	phr
1. SS RMA 1X	100.00
2. Magnesium carbonate	2.50
3. Lime	2.50
4. Zinc oxide	5.00
5. Stearic acid	1.00
6. Sulfur	5.00
7. Formaldehyde–paratoluidine condensation product	0.25
8. Phenyl beta-naphthylamine (PBNA) antioxidant	1.00
Total	**117.25**

Note:
Curing for 5 hours in an autoclave at 140°C is required.
Specific gravity = 1.02; Wallace plasticity = 31; Mooney scorch time (MS 1+3) at
120°C = 20 minutes; Tensile strength = 160 kg/cm^2; Ultimate elongation = 800%;
Modulus at 500% elongation = 30 kg/cm^2; Shore hardness = 36°A.

This is a light gray colored roll compound having a hardness of 35–38°A. For making a black colored compound add 1 phr of fast extrusion furnace (FEF) carbon black. A similar compound is used for making rubber bungs and in rubber linings of pipes.

The accelerator ZMBT (zinc salt of mercaptobenzothiazole (MBT)) can be used in place of MTBZ.

13.4 Cylinder 44 (White)—Paper Mills

Ingredient	phr
1. SS RMA 1X	100.00
2. Burnt lime	2.00
3. Light calcined magnesia (MgO)	2.00
4. Zinc oxide	75.00
5. Stearic acid	3.25
6. Sulfur	3.00
7. Accelerator mercaptobenzimidazole	0.80
8. PBNA	1.00
Total	**187.05**

Note:
Curing for 5 hours at 140°C in an autoclave is suggested.
Specific gravity = 1.49; Wallace plasticity = 25; Mooney scorch time (MS 1+3) at
120°C = 15 minutes; Tensile strength = 145 kg/cm^2; Ultimate elongation = 800%;
Modulus at 500% elongation = 65 kg/cm^2; Shore hardness = 40–45°A.

This is a dull white compound having a hardness of 45°A. For black
coloring add 2 phr of carbon black.

13.5 Cylinder 55—Paper Mills

Ingredient	kg
1. Hevea crumb	42.28
2. Fine China clay	10.15
3. Light calcined MgO	0.85
4. Calcium oxide	0.85
5. Stearic acid	0.42
6. Zinc oxide	42.28
7. MBT accelerator	0.85
8. Carbon black	0.85
9. Sulfur	1.47
Total	**100.00**

Note:
Cure for 4 hours at 140°C.
Specific gravity = 1.65; Wallace plasticity = 30; Mooney scorch time (MS 1+3) at
120°C = 10 minutes; Tensile strength = 140 kg/cm^2; Ultimate elongation = 690%;
Modulus at 500% elongation = 70 kg/cm^2; Shore hardness = 55°A.

Hevea crumb can be replaced with RMA 1X or Pale crepe.

13.6   Cylinder 65—Paper Mills

Ingredient	phr
1. SS RMA 1X	100.00
2. China clay	24.00
3. Light calcined MgO	2.00
4. Burnt lime	2.00
5. Zinc oxide	100.00
6. Stearic acid	1.00
7. Antioxidant PBNA	2.00
8. Sulfur	7.50
Total	**238.50**

Note:
Cure for 5 hours at 140°C.
Specific gravity = 1.69; Wallace plasticity = 30; Mooney scorch time (MS 1+3) at
120°C = 10 minutes; Tensile strength = 130 kg/cm^2; Ultimate elongation = 480%;
Modulus at 200% elongation = 50 kg/cm^2; Shore hardness = 65°A.

For black color add 2 phr of carbon black.

13.7   Cylinder 56 (White)—Paper Mills

Ingredient	kg
1. Heava crumb	50.66
2. Calcium oxide	1.01
3. Light Calcined MgO	1.01
4. Zinc oxide	38.00
5. Titanium dioxide	4.05
6. Stearic acid	1.62
7. Diphenyl guanidine (DPG) accelerator	0.30
8. Sulfur	2.43
9. MTBZ	0.41
10. MBT	0.51
Total	**100.00**

Note:
Cure for 2 hours at 150°C.
Specific gravity = 1.48; Wallace plasticity = 25; Mooney scorch time (MS 1+3) at
120°C = 15 minutes; Tensile strength = 200 kg/cm^2; Ultimate elongation = 660%;
Modulus at 300% elongation = 55 kg/cm^2; Shore hardness = 55°A.

13.8   Cylinder 75—Paper Mills

Ingredient	phr
1. SS RMA 1X	100.00
2. China clay	24.00
3. MgO	2.00
4. Burnt lime	2.00
5. Zinc oxide	100.00
6. Stearic acid	1.00
7. Sulfur	12.00
8. PBNA	2.00
Total	**243.00**

Note:
Cure for 3 hours at 140°C.
Specific gravity = 1.69; Wallace plasticity = 35; Mooney scorch time (MS 1+3) at 120°C = 6 minutes; Tensile strength = 100 kg/cm^2; Ultimate elongation = 260%; Modulus at 100% elongation = 40 kg/cm^2; Shore hardness = 75°A.

For black color add 2 phr of carbon black.

13.9   Cylinder 60—Paper Mills

Ingredient	kg
1. Hevea crumb	41.96
2. Fine China clay	10.07
3. Light calcined MgO	0.84
4. Calcium oxide	0.84
5. Stearic acid	0.42
6. Zinc oxide	41.96
7. MBT	0.84
8. Carbon black	0.84
9. Sulfur	2.23
Total	**100.00**

Note:
Cure for 2 hours at 150°C.
Specific gravity = 1.68; Wallace plasticity =35; Mooney scorch time (MS 1+3) at 120°C = 8 minutes; Tensile strength = 130 kg/cm^2; Ultimate elongation = 600%; Modulus at 500% elongation = 95 kg/cm^2; Shore hardness = 60°A.

13.10   Cylinder 80—Paper Mills

Ingredient	phr
1. SS RMA 1X	100.00
2. China clay	26.00
3. Burnt lime	2.00
4. MgO	2.00
5. Zinc oxide	100.00
6. Stearic acid	1.00
7. Sulfur	14.00
8. PBNA	2.00
9. Carbon black	2.00
Total	**249.00**

Note:
Cure for 2 hours at 150°C.
Specific gravity = 1.69; Wallace plasticity = 30; Mooney scorch time (MS 1+3) at 120°C = 8 minutes; Tensile strength = 85 kg/cm^2; Ultimate elongation = 230%; Modulus at 100% elongation = 45 kg/cm^2; Shore hardness = 80°A.

13.11   Cylinder 92—Paper Mills

Ingredient	phr
1. SS RMA 1X	100.00
2. China clay	24.00
3. MgO	2.00
4. Burnt lime	2.00
5. Zinc oxide	100.00
6. Stearic acid	1.00
7. Sulfur	19.25
8. Antioxidant PBNA	2.00
Total	**250.25**

Note:
Cure for 2 hours at 150°C.
Specific gravity = 1.71; Wallace plasticity = 37; Mooney scorch time (MS 1+3) at 120°C = 6.5 minutes; Tensile strength = 100 kg/cm^2; Ultimate elongation = 230%; Modulus at 100% elongation = 90 kg/cm^2; Shore hardness = 90°A.

For black color add 2 phr of carbon black.

13.12   Cylinder 96—Paper Mills

Ingredient	phr
1. SS RMA 1X	100.00
2. China clay	24.00
3. Burnt lime	2.00
4. MgO	2.00
5. Zinc oxide	100.00
6. Stearic acid	1.00
7. Sulfur	22.00
8. PBNA	2.00
Total	**253.00**

Note:
Cure for 2 hours at 150°C.
Specific gravity = 1.71; Wallace plasticity = 36; Mooney scorch time (MS 1+3) at
120°C = 10 minutes; Tensile strength = 130 kg/cm^2; Ultimate elongation = 140%;
Modulus at 100% elongation = 105 kg/cm^2; Shore hardness = 90°A.

For black color add 2 phr of carbon black.

13.13   Cylinder 995—Semi-ebonite

Ingredient	phr
1. SS RMA 1X	100.00
2. Neoprene W	10.00
3. Ebonite powder	45.00
4. Burnt lime	3.75
5. MgO	3.75
6. MBTS	2.40
7. Sulfur	26.00
Total	**190.90**

Note:
Cure for 2 hours at 150°C.
Specific gravity = 1.43; Wallace plasticity = 30; Mooney scorch time (MS 1+3) at
120°C = 8 minutes; Shore hardness = 90°A.

The addition of neoprene rubber improves tack. Ebonite powder
reduces the speed of exothermic reaction during curing.

13.14   Cylinder for the Steel Industry (Natural Rubber)

Ingredient	I (phr)	II (phr)
1. RMA 1X	100.00	100.00
2. China clay	25.00	20.00
3. Light calcined MgO	2.00	2.00
4. Burnt lime	2.00	2.00
5. Stearic acid	1.00	1.00
6. Zinc oxide	100.00	100.00
7. Nonox D	2.00	2.00
8. Semi-reinforcing furnace (SRF) black	2.00	2.00
9. Sulfur	12.00	7.50
Total	**246.00**	**236.50**
Shore hardness	65–70°A	55–60°A

Note:
Cure for 2 hours at 150°C.
Specific gravity = 1.65 (I), 1.65 (II); Wallace plasticity = 37, 37; Mooney scorch time (MS 1+3) at 120°C = 10 minutes, 10 minutes; Tensile strength = 200 kg/cm^2, 210 kg/cm^2; Ultimate elongation = 250%, 275%; Shore hardness = 70°A, 65°A.

13.15   Cylinder A (Green) for Textile Mills

Ingredient	phr
1. SS RMA 1X	100.00
2. China clay	80.00
3. Burnt lime	2.00
4. MgO	2.00
5. Stearic acid	1.00
6. Zinc oxide	12.00
7. Blue pigment	1.00
8. Carbon black	1.00
9. Sulfur	6.00
10. PBNA antioxidant	1.00
Total	**206.00**

Note:
Cure for 2 hours at 150°C.
Specific gravity = 1.4; Wallace plasticity = 35; Mooney scorch time (MS 1+3) at 120°C = 10 minutes; Tensile strength = 115 kg/cm^2; Ultimate elongation = 720%; Modulus at 400% elongation = 40 kg/cm^2; Shore hardness = 60°A.

13.16   Cylinder E—Textiles

Ingredient	phr
1. SS RMA 1X	100.00
2. China clay	80.00
3. Burnt lime	2.00
4. Calcined MgO	2.00
5. Stearic acid	1.00
6. Zinc oxide	12.00
7. Sulfur	14.50
8. PBNA	1.00
9. Titanium dioxide	0.85
Total	**213.35**

Note:
Cure for 2 hours at 150°C.
Specific gravity = 1.43; Wallace plasticity = 35; Mooney scorch time (MS 1+3) at
120°C = 10 minutes; Tensile strength = 95 kg/cm^2; Ultimate elongation = 240%;
Modulus at 100% elongation = 50 kg/cm^2; Shore hardness = 85°A.

13.17   Cylinder G—Textiles

Ingredient	phr
1. SS RMA 1X	100.00
2. China clay	80.00
3. Burnt lime	2.00
4. MgO	2.00
5. Stearic acid	1.00
6. Zinc oxide	12.00
7. Sulfur	19.25
8. Antioxidant PBNA	1.00
Total	**217.25**

Note:
Cure for 2 hours at 150°C.
Specific gravity = 1.44; Wallace plasticity = 34; Mooney scorch time (MS 1+3) at
120°C = 10 minutes; Tensile strength = 150 kg/cm^2; Ultimate elongation = 170%;
Modulus at 100% elongation = 120 kg/cm^2; Shore hardness = 95°A.

13.18   Cylinder N55

Ingredient	phr
1. Neoprene WRT	100.00
2. Semi-fat brown factice (neoprene grade)	9.00
3. Naphthenic oil	15.00
4. Stearic acid	2.00
5. China clay	30.00
6. Zinc oxide	4.75
7. Light calcined MgO	4.75
8. Carbon black "A"	2.00
9. Silica filler (Vulcasil S)	6.00
Total	**173.50**

Note:

Cure for 2 hours at 150°C.

Specific gravity = 1.44; Wallace plasticity = 15; Mooney scorch time (MS 1+3) at 120°C = 10 minutes; Tensile strength = 90 kg/cm^2; Ultimate elongation = 770%; Modulus at 500% elongation = 45 kg/cm^2; Shore hardness = 51°A.

This is a neoprene-based roll compound that is used where medium oil resistance is required.

13.19   Cylinder N70

Ingredient	phr
1. Neoprene WRT	75.00
2. Neoprene WB	25.00
3. Paraffin oil	10.00
4. Stearic acid	2.50
5. Paraffin wax	1.00
6. Titanium dioxide	8.00
7. Silica (Vulcasil "S")	40.00
8. Zinc oxide	5.00
9. MgO	4.00
Total	**170.50**

Note:

Cure for 2 hours at 150°C.

Specific gravity = 1.45; Wallace plasticity = 23; Mooney scorch time (MS 1+3) at 120°C = 20 minutes; Tensile strength = 100 kg/cm^2; Ultimate elongation = 580%; Modulus at 400% elongation = 45 kg/cm^2; Shore hardness = 62°A.

This is a neoprene roll with medium oil and fuel resistance and good machinability.

13.20   Neoprene N75

Ingredient	phr
1. Neoprene WRT	75.00
2. Neoprene WB	25.00
3. Paraffin oil	6.60
4. Paraffin wax	1.00
5. Stearic acid	2.00
6. Silica filler	38.00
7. Titanium dioxide	8.00
8. Magnesium oxide	4.00
9. Zinc oxide	5.00
10. Sulfur	2.60
Total	**167.20**

Note:
Cure for 2 hours at 150°C.
Specific gravity = 1.48; Wallace plasticity = 30; Mooney scorch time (MS 1+3) at 120°C = 20 minutes; Tensile strength = 100 kg/cm^2; Ultimate elongation = 500%; Modulus at 200% elongation = 35 kg/cm^2; Shore hardness = 68°A.

Add 5 phr of titanium dioxide for making a white colored compound and 2 phr of carbon black for making a black colored compound.

13.21   Cylinder N90

Ingredient	phr
1. Neoprene WRT	77.50
2. Neoprene WB	22.50
3. Napthenic oil (Elasto 541)	3.30
4. Stearic acid	2.00
5. MgO	4.00
6. Silica filler	52.00
7. Titanium dioxide	8.75
8. Paraffin wax	1.25
9. High styrene resin	10.00
10. Zinc oxide	14.75
11. Sulfur	4.10
12. Phenolic resin	3.30
13. Coumarone-indene (CI) resin	5.50
Total	**208.95**

Note:

Cure for 2 hours at 150°C.

Specific gravity = 1.55; Wallace plasticity = 75; Mooney scorch time (MS 1+3) at 120°C = 5 minutes; Tensile strength = 100 kg/cm^2; Ultimate elongation = 300%; Modulus at 200% elongation = 85 kg/cm^2; Shore hardness = 95°A.

13.22 Cylinder P72

Ingredient	phr
1. High nitrile rubber	100.00
2. Dibutyl phthalate	12.00
3. CI resin	12.00
4. FEF carbon black	70.00
5. Stearic acid	1.00
6. Zinc oxide	5.00
7. Salicylic acid	1.00
8. Plasticator FH (aromatic polyether)	6.00
9. Sulfur	2.00
10. MBTS	1.75
11. Vulcalent A retarder (diphenyl nitrosoamine)	1.20
Total	**211.95**

Note:
Cure for 1 hour at 150°C.
Specific gravity = 1.23; Wallace plasticity = 30; Mooney scorch time (MS 1+3) at
120°C = 18 minutes; Tensile strength = 135 kg/cm^2; Ultimate elongation = 560%
Modulus at 300% elongation = 90 kg/cm^2; Shore hardness = 70°A.

Sulfur is to be added first in the mixing cycle to the rubber unlike for
other rubbers.

13.23 Printing Roll Nitrile Based (Oil Resistant Printing Roll)

Ingredient	kg
1. Nitrile rubber (high acrylonitrile¾Krynac 801)	33.30
2. Nitrile rubber extended with plasticizer (Krynac 843)	66.70
3. Stearic acid	0.80
4. Zinc oxide	4.20
5. Thermal black	21.00
6. Dibutyl phthalate	25.00
7. MBTS	1.25
8. Sulfur	1.45
9. Brown factice (synthetic base)	42.00
10. Antioxidant PBNA	0.80
Total	**196.50**

Note:
Cure for 2 hours at 150°C.
Specific gravity = 1.40; Wallace plasticity = 25; Mooney scorch time (MS 1+3) at 120°C = 8 minutes; Tensile strength = 110 kg/cm^2; Ultimate elongation = 450%; Modulus at 300% elongation = 50 kg/cm^2; Shore hardness = 30°A.

13.24 Cylinder "O" for the Textile Industry

Ingredient	kg
1. SMR grade Hevea	49.24
2. Fine clay	39.39
3. Calcium oxide	0.98
4. Light calcined MgO	0.98
5. Stearic acid	0.49
6. Zinc oxide	5.91
7. Sulfur	2.51
8. Antioxidant MC (Nonox NSN) (phenol-aldehyde-amine condensate)	0.50
Total	**100.00**

Note:
Cure for 2 hours at 150°C.
Specific gravity = 1.38; Wallace plasticity = 28; Mooney scorch time (MS 1+3) at 120°C = 10 minutes; Tensile strength = 110 kg/cm^2; Ultimate elongation = 630%; Modulus at 500% = elongation 80 kg/cm^2; Shore hardness = 58°A.

13.25 Cylinder B (Beige) for the Textile Industry

Ingredient	kg
1. RMA 1X	48.22
2. Fine china clay	38.58
3. Calcium oxide	0.97
4. Light calcined MgO	0.97
5. Stearic acid	0.48
6. Zinc oxide	5.79
7. Sulfur	4.45
8. Antioxidant MC	0.48
9. Vulcafor Fast Yellow GTS	0.03
10. Vulcafor Red MS	0.03
Total	**100.00**

Note:
Cure for 2 hours at 150°C.
Specific gravity = 1.41; Wallace plasticity = 38; Mooney scorch time (MS 1+3) at 120°C = 7 minutes; Tensile strength = 70 kg/cm^2; Ultimate elongation = 430%; Modulus at 100% elongation = 25 kg/cm^2; Shore hardness = 68–70°A.

13.26 Cylinder H (Green-Blue) for the Textile Industry

Ingredient	kg
1. RMA 1X	47.89
2. Fine china clay	38.31
3. Calcium oxide	0.96
4. Light calcined MgO	0.96
5. Stearic acid	0.48
6. Zinc oxide	5.75
7. Sulfur	5.08
8. Antioxidant MC	0.48
9. Vulcafor Fast Blue BS	0.09
Total	**100.00**

Note:
Cure for 2 hours at 150°C.
Specific gravity = 1.42; Wallace plasticity = 33; Mooney scorch time (MS 1+3) at 120°C = 7 minutes; Tensile strength = 70 kg/cm^2; Ultimate elongation = 320%; Modulus at 100% elongation = 35 kg/cm^2; Shore hardness = 75°A.

13.27 Cylinder C (Red) for the Textile Industry

Ingredient	kg
1. RMA 1X	47.51
2. Fine china clay	38.00
3. Calcium oxide	0.95
4. Light calcined MgO	0.95
5. Stearic acid	0.48
6. Zinc oxide	5.70
7. Sulfur	5.44
8. Antioxidant MC	0.48
9. Titanium dioxide	0.30
10. Vulcafor Red MS	0.19
Total	**100.00**

Note:
Cure for 2 hours at 150°C.
Specific gravity = 1.43; Wallace plasticity = 29; Mooney scorch time (MS 1+3) at
120°C = 7 minutes; Tensile strength = 90 kg/cm^2; Ultimate elongation = 300%
Modulus at 100% elongation = 45 kg/cm^2; Shore hardness = 80°A.

13.28 Cylinder E (Yellow-Green) for the Textile Industry

Ingredient	kg
1. RMA 1X	46.86
2. Fine china clay	37.49
3. Calcium oxide	0.94
4. Light calcined MgO	0.94
5. Stearic acid	0.47
6. Zinc oxide	5.62
7. Sulfur	6.78
8. Antioxidant MC	0.47
9. Vulcafor Fast Yellow GTS	0.03
10. Titanium dioxide	0.40
Total	**100.00**

Note:
Cure for 2 hours at 150°C.
Specific gravity = 1.43; Wallace plasticity = 38; Mooney scorch time (MS 1+3) at
120°C = 8 minutes; Tensile strength = 95 kg/cm^2; Ultimate elongation = 240%;
Modulus at 100% elongation = 50 kg/cm^2; Shore hardness = 85°A.

13.29   Cylinder F (Light Brown) for the Textile Industry

Ingredient	kg
1. RMA 1X	46.86
2. Fine china clay	37.52
3. Calcium oxide	0.94
4. Light calcined MgO	0.94
5. Stearic acid	0.47
6. Zinc oxide	5.63
7. Sulfur	7.05
8. Antioxidant MC	0.47
9. Vulcafor Fast Yellow GTS	0.03
10. Vulcafor Red MS	0.09
Total	**100.00**

Note:
Cure for 2 hours at 150°C.
Specific gravity = 1.44; Wallace plasticity = 26; Mooney scorch time (MS 1+3) at 120°C = 7 minutes; Tensile strength = 95 kg/cm²; Ultimate elongation = 200%; Modulus at 100% elongation = 55 kg/cm²; Shore hardness = 90°A.

13.30   Cylinder G (Light Green) for the Textile Industry

Ingredient	kg
1. RMA 1X	45.99
2. Fine china clay	36.79
3. Light calcined MgO	0.92
4. Calcium oxide	0.92
5. Stearic acid	0.46
6. Zinc oxide	5.52
7. Sulfur	8.85
8. Antioxidant MC	0.46
9. Vulcafor Emerald Green-A	0.09
Total	**100.00**

Note:
Cure for 2 hours at 150°C.
Specific gravity = 1.44; Wallace plasticity = 34; Mooney scorch time (MS 1+3) at 120°C = 8 minutes; Tensile strength = 150 kg/cm²; Ultimate elongation = 170%; Modulus at 100% elongation = 120 kg/cm²; Shore hardness = 95°A.

13.31   Ethylene-Propylene Diene Monomer Roll for 15% Nitric Acid—Electroplating Service

Ingredient	phr
1. EPCAR 346	100.00
2. General purpose furnace black	110.00
3. Zinc oxide	5.00
4. Paraffin oil	40.00
5. MBTS	2.50
6. LDA	0.80
7. Sulfur	1.50
8. Zinc stearate	0.50
Total	**260.30**

Note:

Cure for 11 hours at 140°C.

Specific gravity = 1.18; Wallace plasticity = 30; Mooney scorch time (MS 1+3) at 120°C = 10 minutes; Tensile strength = 89 kg/cm^2; Ultimate elongation = 320%; Modulus at 300% elongation = 82 kg/cm^2; Shore hardness = 75°A.

13.32 Neoprene Printing Roll—40–45°A

Ingredient	phr
1. Neoprene WHV	40.00
2. Neoprene GRT	60.00
3. Octamine	2.00
4. Stearic acid	1.50
5. Paraffin wax	1.00
6. Petroleum jelly	1.00
7. AC polyethylene	2.00
8. Light calcined MgO	4.00
9. Celite PF3	20.00
10. Fine thermal black	20.00
11. Factice brown	20.00
12. Dutrex R	25.00
13. CI resin	5.00
14. Zinc oxide	5.00
15. NA22	0.20
Total	**206.70**

Note:

Cure for 2.5 hours at 140°C.

Specific gravity = 1.48; Wallace plasticity = 30; Mooney scorch time (MS 1+3) at 120°C = 10 minutes; Tensile strength = 95 kg/cm^2; Ultimate elongation = 600%; Modulus at 100% elongation = 35 kg/cm^2; Shore hardness = 45°A.

13.33   Neoprene Hard Roll Compound (Nonblack)—85°A

Ingredient	phr
1. Neoprene WRT	100.00
2. Octamine	2.00
3. Light calcined MgO	4.00
4. Stearic acid	1.00
5. Hard clay	55.00
6. Precipitated silica	40.00
7. Triethanolamine	3.00
8. AC polyethylene	3.00
9. Petroleum jelly	1.00
10. Process oil	5.00
11. Zinc oxide	5.00
12. NA22	0.50
Total	**219.50**

Note:

Cure in an autoclave for 3 hours at 150°C.

Specific gravity = 1.50; Wallace plasticity = 40; Mooney scorch time (MS 1+3) at 120°C = 10 minutes; Tensile strength = 100 kg/cm^2; Ultimate elongation = 510%; Modulus at 300% elongation = 55 kg/cm^2; Shore hardness = 85°A.

13.34   Hypalon Roll (Black)—85°A

Ingredient	phr
1. Hypalon 40	100.00
2. Epikote resin 828	15.00
3. MBTS	0.50
4. Tetrone A	1.25
5. DOTG	0.25
6. NBC	1.00
7. SRF	60.00
8. Koresin	5.00
9. Dioctylphthalate	5.00
10. AC polyethylene	3.00
Total	**191.00**

Note:
Cure for 2.5–3 hours at 150°C in an autoclave.
Specific gravity = 1.45; Wallace plasticity = not consistent; Mooney scorch time
(MS 1+3) at 120°C = 8 minutes; Tensile strength = 140 kg/cm²;
Ultimate elongation = 380%; Modulus at 200% elongation = 85 kg/cm²; Shore
hardness = 80°A.

Calendering is easier with this compound because of the presence of adequate plasticizer levels.

Building tack, which is normally poor in Hypalon compounds, is improved with the addition of Epikote resin.

Local heating of the calendered sheet may be required while building up the roll.

13.35   Hypalon Roll Compound (White)—98°A

Ingredient	phr
1. Hypalon 30	100.00
2. Light calcined MgO	20.00
3. AC polyethylene	3.00
4. Tipure R610 (titanium dioxide)	35.00
5. Hisil	25.00
6. Epikote resin 828	20.00
7. Tetrone A	2.00
8. Phthalic anhydride	13.00
Total	**218.00**

Note:
Cure for 30 minutes at 150°C.
Specific gravity = 1.40.

This is a test compound suggested for special Hypalon rolls. The proportion of silica filler and resin can be adjusted for ease of processing.

13.36   Rubber Roll for Tannery—60°A

Ingredient	phr
1. RMA 1X	100.00
2. Zinc oxide	5.00
3. Nonox SP	1.00
4. Nonox WSP	0.50
5. Stearic acid	1.00
6. Ultrasil	42.00
7. Diethylene glycol (DEG)	1.00
8. Flexon 840	5.00
9. Rosin	1.00
10. Sulfur	5.00
11. Ultra blue	0.30
12. Titanox	5.00
13. DPG	0.50
14. Vulcacit CZ	0.80
Total	**168.10**

Note:
Cure for 2 hours at 153°C in an autoclave.
Specific gravity = 1.30; Wallace plasticity = 30; Mooney scorch time (MS 1+3)
at 120°C = 10 minutes; Tensile strength = 140 kg/cm^2; Ultimate elongation = 500%;
Modulus at 300% elongation = 65 kg/cm^2; Shore hardness = 60°A.

13.37   Rubber Roll for Tannery—80°A

Ingredient	phr
1. RMA 1X	100.00
2. Zinc oxide	5.00
3. Vulcasil S	92.50
4. Petroleum jelly	2.00
5. Nonox SP	1.00
6. DEG	3.00
7. Elasto 710	7.50
8. Stearic acid	2.50
9. Santocure CBS	0.75
10. Sulfur	4.00
Total	**218.25**

Note:

Cure for 2 hours at 150°C in an autoclave.

Specific gravity = 1.30; Wallace plasticity = 32; Mooney scorch time (MS 1+3) at 120°C = 10 minutes; Tensile strength = 170 kg/cm^2; Ultimate elongation = 450%; Modulus at 300% elongation = 90 kg/cm^2; Shore hardness = 80°A.

14 Tank Linings and Adhesives

14.1 Introduction

Rubber is used in corrosion-proof linings, more than any other material, because of its proven superiority in such applications. Fertilizer, steel plants, chemical, caustic soda, paper, ore, mining, and pharmaceutical industries use rubber linings in process equipment where corrosive media must be handled. Synthetic and natural rubbers form a unique class of high polymers having worldwide acceptance in combating corrosion. One of the most important characteristics of natural and man-made rubbers is their remarkable resistance against corrosive chemicals, fumes, acids, and alkaline and salt solutions handled in chemical plants and equipments such as storage and reaction tanks, ducts, vessels, pipings, and so on. In the absence of anticorrosive rubber linings, corrosion would be so extensive that most products of modern technology could not exist.

The compound formulation technique for producing a useful end product for application in anticorrosion linings and coatings calls for thorough knowledge of rubbers and the chemicals which are admixed with them. The compounds are calendered to form a uniform sheet of thickness ranging from 2 to 10 mm, as per requirements. The most commonly adopted thickness is 5 mm for corrosion resistance and 10 mm for abrasion resistance. The metal surface is cleaned by sand/shot blasting, applied with a coat of adhesive and then the rubber sheets are laid/lined over the coated surface. The lined equipment is then cured/vulcanized in an autoclave under specified temperature and pressure conditions. Various curing methods are adopted such as autoclave curing, hot water curing, open steam curing, and cold bonding, with prevulcanized sheets.

Some rubber lining formulations for various duty conditions are given below.

Note:

1. For ease of incorporation during the mixing operation in the mixing mill, sulfur is added in the form of sulfur master batch (SMB) which is made as follows:

Ingredient	phr	Batch Weight (kg)
1. SS RMA 1X	100.00	8
2. Sulfur	200.00	16
Total	**300.00**	**24**

Sulfur is mixed well with rubber and the master batch is used as an ingredient in many lining formulations as required.

2. The proportions in the rubber lining formulations are given for a batch capacity of a 22″ × 60″ rubber mixing mill. The batch size has to be reduced for smaller mills.

3. Premastication of raw rubber is required wherever a Mooney viscosity of 30 is specified. A Mooney viscometer is used to measure the viscosity of the rubber. From a practical point of view, premastication for 30–40 minutes will give masticated rubber with a Mooney viscosity of 30–40 units or premastication for 15–20 minutes with a peptizing agent will also give masticated rubber with a Mooney viscosity of 30–35 units.

4. Curing of rubber-lined vessels in the autoclave is done at the curing temperature in steps starting from the room temperature.

14.2 Rubber Lining of Digesters with Brick Lining for Sulfuric Acid Conditions at 100°C

Ingredient	kg
1. Butyl rubber (Polysar 300)	20.000
2. Paraffin wax	0.400
3. China clay	9.352
4. Talc powder	6.552
5. Zinc oxide	1.000
6. Magnesium oxide	1.000
7. Stearic acid	0.180
8. Fast extrusion furnace (FEF) black	2.060
9. MBTS	0.150
10. Tetramethylthiuram disulfide (TMT)	0.300
11. Sulfur	0.400
Batch weight	**41.394**

Note:
Specific gravity = 1.36; Wallace plasticity = 42; Mooney scorch time (MS 1+3) at 120°C = 17 minutes; Tensile strength = 110 kg/cm²; Ultimate elongation = 640%; Modulus at 300% elongation = 40 kg/cm²; Shore hardness = 60°A.

A 5 mm thick rubber sheet is cured in an autoclave for 4 hours at 140°C and used for lining with a cold bond adhesive.

14.3 Rubber Lining Drum Filters for Handling Sulfuric Acid Slurry at 65°C

Ingredient	kg
1. Hypalon 40	16.000
2. Hypalon 20	4.000
3. Antioxidant NBC	0.600
4. Tetrone A	0.200
5. MBTS	0.100
6. China clay	8.000
7. Litharge	5.000
8. Epoxy resin	3.000
9. Aromatic process oil	1.000
10. Plasticator FH (aromatic polyether)	1.000
11. Low density polyethylene	0.800
Total	**39.700**

Note:
Specific gravity = 1.50; Wallace plasticity = 45; Mooney scorch time (MS 1+3) at
120°C = 8 minutes; Tensile strength = 98 kg/cm^2; Ultimate elongation = 400%;
Modulus at 300% elongation = 44 kg/cm^2; Shore hardness = 65°A.

The rubber-lined equipment is cured in an autoclave for 5 hours at 140°C.

14.4 Rubber Lining for Iron Ore Slurry (Wear Resistant)

Ingredient	kg
1. RMA 1X (30 Mooney)	30.000
2. Aromatic process oil	0.750
3. Semi-reinforcing furnace (SRF) carbon black	6.000
4. Zinc oxide	3.000
5. Stearic acid	0.600
6. Litharge	1.800
7. Mercaptobenzothiazole (MBT)	0.060
8. Phenyl beta-naphthylamine (PBNA)	0.300
9. Sulfur	0.960
Total	**43.470**

Note:
Cure for 2 hours at 140°C.
Specific gravity = 1.15; Wallace plasticity = 30; Mooney scorch time (MS 1+3) at
120°C = 8 minutes; Tensile strength = 200 kg/cm^2; Ultimate elongation = 600%;
Modulus at 300% elongation = 85 kg/cm^2; Shore hardness = 45°A.

The stock is mixed without sulfur which is to be added at the time of calendering.

14.5 Adhesive Solution for the Abrasion/Wear Resistant Lining Compound for Slurry Lines

Ingredient	kg
1. RMA 1X	10.000
2. Zinc oxide	0.500
3. Stearic acid	0.100
4. Paraffin wax	0.300
5. Naphthenic oil	0.600
6. SRF black	7.400
7. HBS	0.120
8. Sulfur	0.230
9. Salicylic acid	0.030
10. PBNA	0.100
11. Coumarone-indene (CI) resin	1.500
Total	**20.880**

Note:
Specific gravity = 1.20; Wallace plasticity = 30; Mooney scorch time (MS 1+3) at 120°C = 8 minutes; Tensile strength = 175 kg/cm^2; Ultimate elongation = 550%; Modulus at 300% elongation = 70 kg/cm^2; Shore hardness = 55°A.

The compound is mixed and dissolved in toluene in a 75 : 25 ratio and used as a secondary coating on the metal surface and then the lining compound is laid. The primary coating will be any isocyanate-based proprietary bonding agent such as Chemlok. The lined equipment/pipe is cured at 140°C for 4 hours in an autoclave.

14.6 Rubber Lining for Wet Chlorine—Caustic Soda Industry

The base layer is a semi-ebonite compound usually 2 mm thick and the lining layer is a full ebonite compound usually 3 mm thick.

14.6.1   Base Layer (Semi-ebonite)

Ingredient	kg
1. RMA 1X	23.50
2. Neoprene WB	3.00
3. Naphthenic Oil	0.600
4. CBS accelerator	0.600
5. Talc	13.796
6. SMB	10.574
Total	**52.070**

Note:
Specific gravity = 1.44; Shore hardness = 60°D.

14.6.2   Lining Layer (True Ebonite)

Ingredient	kg
1. RMA 1X (30 Mooney)	21.816
2. Ebonite powder	3.312
3. Burnt lime	0.504
4. High abrasion furnace (HAF) black	0.612
5. Graphite powder	6.048
6. Salicylic acid	0.186
7. Accelerator F	0.612
8. Antioxidant PBNA	0.504
9. SMB	14.568
Total	**48.162**

Note:
Specific gravity = 1.30; Shore hardness = 80°D.

   Both compounds are separately calendered and doubled to a total thickness of 5 mm and then used for lining. The lined equipment is cured in an autoclave for 10 hours at 140°C in steps.

14.7 Adhesive Dissolution to be Used on Cleaned Metal Surfaces for Ebonite Lining

Ingredient	kg
1. SS RMA 1X	14.286
2. Calcium carbonate	2.856
3. Sulfur	2.856
4. Zinc oxide	20.000
5. MBTS	0.442
Total	**40.440**

Note:
Specific gravity = 1.75; Wallace plasticity = 28; Mooney scorch time (MS 1+3) at 120°C = 12 minutes; Tensile strength = 45 kg/cm^2; Ultimate elongation = 150%; Shore hardness = 75–80°A.

The compound is mixed in a mixing mill and dissolved in either trichloroethylene or hexane (120 liters for 40.44 kg) and the dissolution is done for 2.5 hours in a deflocculator to homogenize the mixture.

14.8 Lining Formulations for Phosphoric Acid Storage Tanks

Two types of linings based on natural and neoprene rubbers are applied by the cold bonding technique using a neoprene-based adhesive bonding solution. Storage tanks at site terminals are to be inspected for their suitability for lining such as having a smooth, air-free welded surface. The tank is sand blasted and then the adhesive coating is applied over which the prevulcanized acid resistant sheet is lined.

14.8.1   Natural Rubber

Ingredient	kg
1. RMA 1X (30 Mooney)	30.000
2. Aromatic process oil	0.750
3. SRF black	12.000
4. Zinc oxide	3.000
5. Stearic acid	0.600
6. Litharge	1.800
7. MBT	0.060
8. PBNA	0.300
9. Sulfur	0.960
Total	**49.470**

Note:
Specific gravity = 1.20; Wallace plasticity = 25; Mooney scorch time (MS 1+3) at 120°C = 7 minutes; Tensile strength = 200 kg/cm^2; Ultimate elongation = 575%; Modulus at 300% elongation = 135 kg/cm^2; Shore hardness = 52°A.

Sulfur is to be added at the time of calendering the compound.

14.8.2   Neoprene Rubber

Ingredient	kg
1. Neoprene WRT	15.150
2. Neoprene WB	6.492
3. China clay	7.580
4. Talc powder	4.330
5. Naphthenic Oil	0.866
6. Zinc oxide	0.650
7. Magnesium oxide	0.866
8. Low density polyethylene	0.400
9. MBTS	0.014
10. TMT	0.172
11. SRF black	2.596
12. PBNA	0.432
13. Sulfur	0.216
Total	**39.890**

Note:
Specific gravity = 1.55; Wallace plasticity = 32; Mooney scorch time (MS 1+3) at 120°C = 20 minutes; Tensile strength = 145 kg/cm^2; Ultimate elongation = 645%; Modulus at 300% elongation = 70 kg/cm^2; Shore hardness = 65–68°A.

14.9   Cold Bond Adhesive Common for Natural, Neoprene, Butyl, and Hypalon Rubber Sheets

Ingredient	kg
1. Neoprene AC	14.000
2. Neoprene WB	4.000
3. Toluene	17.3000
4. Zinc oxide	1.000
5. Magnesium oxide	1.000
6. Phenol-formaldehyde resin	2.000
7. Chlorinated rubber	3.000
8. Ethyl acetate	39.400
9. Hexane	26.000
10. Acetone	0.635
11. Nonox NSN	1.000
12. Fine silica powder	2.000
Total	**110.435**

The adhesive is made in a "Z" blade mixing equipment to the required consistency. Before applying on a cleaned metal surface, 1% of an iso-cyanate bonding agent such as Desmodur-R is to be mixed with the adhesive as required.

14.10   Mixture of Solvents for Evaporation Makeup

Ingredient	kg	Liters
1. Toluene	17.000	20.00
2. Ethyl acetate	40.020	46.000
3. Hexane	21.580	33.200
4. Acetone	0.620	0.800
Total	**79.220**	**100.000**

14.11 Chlorine Resistant Compound Formulation Used in Mercury Cells in the Caustic Soda Industry

Ingredient	kg
1. RMA 1X (30 Mooney)	22.000
2. Zinc oxide	1.100
3. Stearic acid	0.660
4. FEF black	6.600
5. SRF black	6.600
6. Graphite powder	6.600
7. Elasto 710 process oil	2.200
8. Vulcacit CZ	0.260
9. Sulfur	0.440
10. Nonox HFN	0.180
Total	**46.640**

Note:
Specific gravity = 1.40; Wallace plasticity = 35; Mooney scorch time (MS 1+3) at 120°C = 10 minutes; Tensile strength = 120 kg/cm^2; Ultimate elongation = 400%; Modulus at 300% elongation = 90 kg/cm^2; Shore hardness = 65°A.

This compound (hardness = 55–60°A) is mixed and calendered to a thickness of 3 mm and doubled with a 2 mm neoprene sheet (hardness = 50–55°A) on top, wound on drums and then cured in an autoclave for 3 hours at 141°C.

14.12   Semi-ebonite Compound Formulation for Extrusion of Profiles for Drum Filters

Ingredient	kg
1. RMA 1X	20.000
2. Zinc oxide	1.000
3. Stearic acid	0.600
4. FEF black	12.000
5. China clay	5.000
6. SRF black	4.000
7. Naphthenic oil	1.000
8. Vulcacit CZ	0.240
9. Sulfur	3.000
Total	**46.840**

Note:
Specific gravity = 1.30; Wallace plasticity = 40; Mooney scorch time (MS 1+3)
at 120°C = 12 minutes; Shore hardness = 85°A.

This is a flexible ebonite compound for extrusion of profiles such as division strips, end rings, and gaskets for use in drum filters in the chemical industry.

14.13   Formulation for Sulfuric Acid/Chlorine Solutions in Drying Towers in the Caustic Soda Industry

Ingredient	kg
1. Hypalon 40	12.000
2. Hypalon 20	8.000
3. Antioxidant NBC	0.600
4. Tetrone A accelerator	0.200
5. MBTS	0.100
6. Fine china clay	8.000
7. Litharge	5.000
8. Epoxy resin (Gy250)	3.000
9. Aromatic process oil	1.000
10. Plasticator FH	1.000
11. Low density polyethylene	0.800
Total	**39.700**

Note:
Specific gravity = 1.50; Wallace plasticity = 45; Mooney scorch time (MS 1+3) at 120°C = 10 minutes; Tensile strength = 98 kg/cm^2; Ultimate elongation = 420%; Modulus at 300% elongation = 45 kg/cm^2; Shore hardness = 65°A.

Calendering of this compound is a tricky job and involves special skills. Warming in mills and temperature controls should be done carefully.

14.14 Ebonite Based on Styrene Butadiene Rubber (SBR) for Making Ebonite Distance Pieces and Internals for the Chemical Industry

Ingredient	kg
1. SBR (Synaprene 1502)	18.000
2. High styrene resin	9.000
3. FEF black	10.800
4. HAF black	3.600
5. Aromatic process oil	0.900
6. Zinc oxide	0.900
7. Stearic acid	0.360
8. HBS accelerator	0.234
9. Sulfur	2.700
10. Plasticator FH	0.360
11. CI resin	1.200
Total	**48.054**

Note:
Specific gravity = 1.44; Wallace plasticity = 40; Mooney scorch time (MS 1+3) at 120°C = 10 minutes; Shore hardness = 90°A (70°D).

Tubular components are either extruded or hand-built on mandrels using the above compound, wrapped with thick cotton duck, and cured in hot water. After curing they are ground to a smooth finish. Hot water temperature is 100°C and curing time is 24 hours.

14.15 Ebonite Formulation Suitable for the Hot Water Curing Method

Ingredient	kg
1. RMA 1X (30 Mooney)	16.200
2. Ebonite powder	6.000
3. Burnt lime	0.500
4. SRF black	0.500
5. PBNA	0.400
6. Zinc oxide	7.600
7. Graphite powder	5.000
8. Salicylic acid	0.360
9. SMB	11.400
Total	**47.96**
10. Accelerator TMT	1.000
11. Accelerator LDA	0.100
Total	**49.060**

Note:
Specific gravity = 1.40; Wallace plasticity = 28; Mooney scorch time (MS 1+3) at 120°C = 10 minutes; Shore hardness = 98°A (70°D).

While calendering add the accelerators during the warming stage. This compound has high proportions of zinc oxide. The lined tank is filled with water heated with steam for curing.

14.16 Acid Resistant Strip Extrusion Compound Formulation (Natural Rubber)

Ingredient	kg
1. RMA 1X	22.22
2. Zinc oxide	1.11
3. Stearic acid	0.64
4. FEF black	4.44
5. SRF black	10.00
6. Process oil (Elasto 710)	0.66
7. MBT	0.10
8. Vulcalent A (retarder)	0.10
9. Sulfur	0.61
10. PBNA	0.22
Total	**40.10**

Note:
Cure for 60 minutes at 140°C.
Specific gravity = 1.20; Wallace plasticity = 29; Mooney scorch time (MS 1+3) at 120°C = 10 minutes; Tensile strength = 150 kg/cm^2; Ultimate elongation = 450%; Modulus at 100% elongation = 65 kg/cm^2; Shore hardness = 65°A.

Vulcalent A is added to reduce scorching tendency during extrusion of profiles.

14.17   Acid and Ozone Resistant Strip Extrusion Compound Formulation (Neoprene Rubber)

Ingredient		kg
1.	Neoprene WRT	7.90
2.	Neoprene WB	3.38
3.	FEF black	4.50
4.	Process oil (Elasto 710)	1.69
5.	Brown factice	1.13
6.	Paraffin wax	0.22
7.	Magnesium oxide	0.44
8.	Zinc oxide	0.57
9.	Vulcacit NPV	0.22
10.	MBTS	0.11
	Total	**20.160**

Note:
Cure for 60 minutes at 140°C.
Specific gravity = 1.55; Wallace plasticity = 30; Mooney scorch time (MS 1+3) at 120°C = 15 minutes; Tensile strength = 150 kg/cm^2; Ultimate elongation = 650%; Modulus at 300% elongation = 70 kg/cm^2; Shore hardness = 68°A.

Brown factice is added for ease of extrudability.

14.18 Nitric Acid Resistant Ethylene-Propylene Diene Monomer (EPDM) Lining Formulation for the Electroplating Industry

	Ingredient	kg
1.	EPDM rubber	5.00
2.	Fine china clay	5.00
3.	Paraffin oil	3.50
4.	CI resin or phenolic resin	0.50
5.	Zinc oxide	0.25
6.	Paraffin oil (Flexon 840)	1.50
7.	Stearic acid	0.15
8.	LDA	0.04
9.	Thiuram	0.04
10.	Tetrone A	0.04
11.	MBT	0.08
12.	Sulfur	0.10
13.	HSL	0.10
	Total	**16.30**

Note:

Cure for 5 hours at 140°C.

Specific gravity = 1.20; Wallace plasticity = 30; Mooney scorch time (MS 1+3) at 120°C = 12 minutes; Tensile strength = 90 kg/cm^2; Ultimate elongation = 350%; Modulus at 300% elongation = 85 kg/cm^2; Shore hardness = 75°A.

14.19   Bromobutyl Lining Formulation for the Ore/Sand Beneficiation Industry

Ingredient	kg
1. Bormobutyl rubber	26.000
2. FEF black	13.000
3. Zinc oxide	1.040
4. Paraffin wax	0.260
5. Stearic acid	0.260
6. TMT accelerator	0.130
7. MBTS	0.260
8. Sulfur	0.130
Total	**41.080**

Note:
Cure for 4 hours at 140°C.
Specific gravity = 1.21; Wallace plasticity = 29; Mooney scorch time (MS 1+3) at 120°C = 12 minutes; Tensile strength = 88 kg/cm^2; Ultimate elongation = 700%; Modulus at 300% elongation = 20 kg/cm^2; Shore hardness = 58°A.

This compound is very tough for mixing and calendering. Proper temperature and nip control are needed during calendaring.

14.20 White Natural Rubber Compound Formula for Lining of Equipments in Pigment Plants

Ingredient	kg
1. RMA 1X (30 Mooney)	30.000
2. MBTS	0.300
3. PBNA	0.300
4. Zinc oxide	1.500
5. Stearic acid	0.450
6. Elasto 710 oil	0.900
7. Titanium dioxide	6.000
8. Fine talc powder	6.000
9. Sulfur	0.900
10. TMT	0.030
Total	46.380

Note:
Cure for 2 hours at 140°C.
Specific gravity = 1.32; Wallace plasticity = 32; Mooney scorch time (MS 1+3) at 120°C = 12 minutes; Tensile strength = 120 kg/cm^2; Ultimate elongation = 600%; Modulus at 300% elongation = 55 kg/cm^2; Shore hardness = 55°A.

Care should be taken to maintain proper temperature differentials during mixing and calendering to avoid porosity, as this compound is prone to porosity. Low temperature mixing and calendering is preferred.

14.21 White Neoprene Rubber Lining for Pigmentation Plants

Ingredient	phr
1. Neoprene WRT	85.00
2. Neoprene WB	15.00
3. Stearic acid	0.50
4. PBNA	2.00
5. Magnesium oxide	4.00
6. Zinc oxide	5.00
7. Titanium dioxide	15.00
8. Thiuram	0.50
9. Sulfur	1.00
10. DM/C	0.20
11. Vulcacit NPV	1.00
12. Elasto 541	5.00
Total	**134.20**

Note:
Sheets are cured in an autoclave on drums for 4 hours at 130°C.
Specific gravity = 1.40; Wallace plasticity = 28; Mooney scorch time (MS 1+3) at
120°C = 15 minutes; Tensile strength = 174 kg/cm^2; Ultimate elongation = 850%;
Modulus at 300% elongation = 49 kg/cm^2; Shore hardness = 45–50°A.

14.22   White Natural Rubber/Neoprene Blend for Pigmentation Plants

Ingredient	phr
1. RMA 1X	75.00
2. Neoprene WRT	21.25
3. Neoprene WB	3.75
4. DM/C	1.00
5. PBNA	1.00
6. Zinc oxide	5.00
7. Magnesium oxide	1.00
8. Stearic acid	1.50
9. Elasto 710	4.00
10. Titanium dioxide	15.00
11. Talc	20.00
12. Sulfur	2.00
13. Thiuram	0.25
14. Vulcacit NPV	0.25
15. Salicylic acid	0.25
Total	**151.25**

Note:
Cure in an autoclave for 4 hours at 130°C.
Specific gravity = 1.44; Wallace plasticity = 30; Mooney scorch time (MS 1+3) at 120°C = 20 minutes; Tensile strength = 170 kg/cm^2; Ultimate elongation = 650%; Modulus at 300% elongation = 65 kg/cm^2; Shore hardness = 45–50°A.

14.23   Chlorobutyl Adhesive Formulation for Use in Butyl Lining Formulation

Ingredient	kg
1. Chlorobutyl rubber	3.00
2. HAF black	1.20
3. Phenolic resin	0.30
4. Zinc oxide	0.30
5. CI resin	0.60
Total	**5.40**

Dissolve in xylene or toluene in a 75 : 25 ratio and use as a top coat.

14.24   Adhesive for Patchwork in a Rubber-Lined Pipes

14.24.1   Compound A

Ingredient	phr
1. SS RMA 1X	100.00
2. Calcium carbonate	20.00
3. Zinc oxide	20.00
4. MBTS	1.00
5. Sulfur	20.00
6. Nonox D	2.00
Total	**163.00**

14.24.2   Compound B

Ingredient	phr
1. RMA 1X	100.00
2. LDA	2.00
3. CBS	0.50
Total	**102.50**

Dissolve A and B separately in trichloroethylene and mix in a 50 : 50 ratio, stir well, and then apply on the surface to be repaired.

14.25 Butyl Rubber Lining for Acid Regeneration Duty

Acid regeneration duty in a chemical plant generally involves the use of ferric and ferrous chlorides with hydrochloric acid at 80°C.

Ingredient	phr
1. Polysar Butyl 301	100.00
2. Stearic acid	1.00
3. Zinc oxide	10.00
4. FEF black	50.00
5. Paraffin wax	1.00
6. Vulcacit thiuram	2.00
7. DM/C	0.75
8. Sulfur	1.50
Total	**166.25**

Note:
Cure in an autoclave for 3.5–4 hours at 130°C.
Specific gravity = 1.40; Wallace plasticity = 45; Mooney scorch time (MS 1+3) at 120°C = 15 minutes; Tensile strength = 135 kg/cm^2; Ultimate elongation = 600%; Modulus at 300% elongation = 50 kg/cm^2; Shore hardness = 65°A.

14.26 Flexible Cell Covers

These cell covers are used in cell houses in caustic soda plants. These consist of two layers: a 3 mm thick bottom layer and a 2 mm thick top layer. The top layer is neoprene based for compatibility in ozone environments and the bottom layer is resistant to wet chlorine.

14.26.1 Ozone Resistant Neoprene Layer

Ingredient	phr
1. Bayprene 110	100.00
2. Antioxidant HSL	2.00
3. Magnesium oxide	4.00
4. SRF black	35.00
5. China clay	65.00
6. Elasto 541	10.00
7. Paraffin wax	1.00
8. Zinc oxide	5.00
9. TMT	1.00
10. Sulfur	1.00
Total	**224.00**

Note:
Specific gravity = 1.50; Wallace plasticity = 32; Mooney scorch time (MS 1+3) at 120°C = 15 minutes; Tensile strength = 148 kg/cm^2; Ultimate elongation = 650%; Modulus at 300% elongation = 70 kg/cm^2; Shore hardness = 65°A.

14.26.2 Wet Chlorine Resistant Layer

Ingredient	Alternate I (phr)	Alternate II (phr)
1. Natural rubber RMA 1X	100.00	100.00
2. Zinc oxide	5.00	5.00
3. Stearic acid	3.00	3.00
4. Plumbogine or graphite	30.00	30.00
5. China clay	50.00	—
6. FEF black	30.00	30.00
7. SRF black	—	30.00
8. Elasto 710 oil	6.00	10.00
9. Vulcacit CZ	1.20	1.20
10. Sulfur	2.00	2.00
11. Nonox HFN	1.00	1.00
Total	**228.20**	**212.20**

Note:
Specific gravity = 1.40; Wallace plasticity = 30; Mooney scorch time (MS 1+3) at 120°C = 10 minutes; Tensile strength = 130 kg/cm^2; Ultimate elongation = 450%; Modulus at 300% elongation = 85 kg/cm^2; Shore hardness = 65°A.

Both neoprene and natural rubber compounds are calendered separately and doubled, then they are wound on drums with wet cotton duck backing and cured at 130°C for 4 hours in an autoclave.

14.27 Butyl Rubber/EPDM Membrane for Use in Fishery Tanks

Ingredient	phr
1. Polysar 301	65.00
2. Nordel (EPDM)	35.00
3. Stearic acid	1.00
4. Zinc oxide	10.00
5. HAF black	50.00
6. FEF black	20.00
7. Talc	40.00
8. Elasto 541 oil	20.00
9. Antioxidant NBC	1.00
10. Paraffin wax	1.50
11. Sulfur	1.00
12. MBT	2.00
13. MBTS	1.50
Total	**248.00**

Note:
Specific gravity = 1.25; Wallace plasticity = 48; Mooney scorch time (MS 1+3) at 120°C = 15 minutes; Tensile strength = 125 kg/cm^2; Ultimate elongation = 500%; Modulus at 300% elongation = 45 kg/cm^2; Shore hardness = 60°A.

Sheets of required thickness are calendered, wound on drums with cloth backing, and cured in an autoclave for 2.5 hours at 140°C.

14.28   Low Temperature Curable Bromobutyl Lining for Digesters in the Ore Industry

Ingredient	Alternate I (phr)	Alternate II (phr)
1. Polysar bromobutyl	100.00	100.00
2. Medium thermal (MT) black	100.00	100.00
3. Flexon 840	10.00	10.00
4. Low density polyethylene	5.00	5.00
5. Stearic acid	1.00	1.00
6. Red lead	10.00	10.00
7. Vulcacit F	2.00	—
8. TMT	2.00	—
9. Hexamethylenetetramine	—	2.00
10. Ethylene thiourea	—	2.00
Total	**230.00**	**230.00**

Note:
Specific gravity 1.22 (alternate I), 1.22 (alternate II); Wallace plasticity = 40, 40;
Mooney scorch time (MS 1+3) at 120°C = 14 minutes, 10 minutes;
Tensile strength = 88 kg/cm^2, 90 kg/cm^2; Ultimate elongation = 650%, 700%; Modulus
at 300% elongation = 20 kg/cm^2, 25 kg/cm^2; Shore hardness = 55°A, 55°A.

An ultra accelerator system is adopted in both compounds. As these are low temperature curable compounds, care should be taken to keep mixing and calendering temperatures low.

14.29 Open Steam Curable Phosphoric Acid Resistant Soft Natural Rubber Compound

14.29.1 Master Batch

Ingredient	phr
1. RMA 1X (30 Mooney)	100.00
2. Elasto 710 oil	2.50
3. SRF black	40.00
4. Zinc oxide	10.00
5. Stearic acid	2.00
6. Litharge	6.00
7. MBT	0.20
8. Nonox D	1.00
Total	**161.70**

14.29.2 Ultra Accelerator System

Ingredient	phr
1. TMT	5.00
2. Sulfur	3.00
3. Dicup 40	4.00
Total	**12.00**

Note:
Specific gravity = 1.20; Wallace plasticity = 30; Mooney scorch time (MS 1+3) at 120°C = 7 minutes; Tensile strength = 225 kg/cm^2; Ultimate elongation = 590%; Modulus at 300% elongation = 130gs/cm^2; Shore hardness = 54°A.

The ultra accelerator system has to be added at the time of sheeting from the mill or calender.

14.30   Rubber Lining for 20% Hydrochloric Acid at 100°C

Ingredient	phr
1. Nordel (EPDM)	100.00
2. China clay	75.00
3. SRF black	50.00
4. Phenolic resin	20.00
5. Zinc oxide	10.00
6. Flexon 840	10.00
7. Stearic acid	1.00
8. LDA	1.00
9. TMT	2.00
10. MBT	1.00
11. Sulfur	1.75
12. HSL	2.00
Total	**273.75**

Note:
Sheets are cured in an autoclave at 125°C for 3 hours.
Specific gravity = 1.18; Wallace plasticity = 30; Mooney scorch time (MS 1+3) at 120°C = 14 minutes; Tensile strength = 75 kg/cm^2; Ultimate elongation = 330%; Modulus at 300% elongation = 62 kg/cm^2; Shore hardness = 68–70°A.

At lower temperatures of up to 75–80°C natural rubber ebonite is resistant to hydrochloric acid.

14.31   Lining of Impellers in Phosphatic Fertilizer Plants with Tip Velocities of 109–110 m/sec to withstand Dust and Fumes of Phosphoric Acid

Ingredient	Batch Weight (kg)
1. SS RMA 1X	24.00
2. Neoprene WB	3.00
3. Elasto 641	0.60
4. CBS	0.60
5. Fine talc	10.00
6. SMB	10.50
7. FEF black	5.00
Total	**53.70**

Note:
Cure for 3 hours at 140°C in an autoclave.
Specific gravity = 1.45; Wallace plasticity = 25; Mooney scorch time (MS 1+3) at 120°C = 14 minutes; Shore hardness = 90°A.

Calender to the required thickness of 3 or 5 mm and then use the compound for lining the impellers.

14.32   Low Water Absorption Neoprene Lining Formulation for Chlor-Alkali Plants

The cell boxes in mercury cells in chlor-alkali plants require a low water absorption rubber lining. In such cases, in addition to the ebonite lining, a neoprene-based lining with low water absorption and ozone resistance is applied over it.

Ingredient	phr
1. Neoprene WRT	70.00
2. Neoprene WB	30.00
3. AC polythene	3.00
4. China clay	35.00
5. Talc	20.00
6. Elasto 710	4.00
7. Red lead	20.00
8. Sulfur	1.00
9. Monosulfide	0.80
10. SRF black	12.00
11. Paraffin wax	2.00
Total	**197.80**

Note:
Cure in an autoclave for 4 hours at 130°C.
Specific gravity = 1.65; Wallace plasticity = 38; Mooney scorch time (MS 1+3) at
120°C = 20 minutes; Tensile strength = 140 kg/cm^2; Ultimate elongation = 640%;
Modulus at 300% elongation = 70 kg/cm^2; Shore hardness = 65–68°A.

Red lead gives high water and water vapor resistance.

14.33 Butyl Lining for Digester (Without Mineral Fillers)

Ingredient	Batch Weight (kg)
1. Polysar 301	24.00
2. Stearic acid	0.24
3. Zinc oxide	2.40
4. FEF black	6.00
5. MT black	6.00
6. Paraffin wax	0.24
7. Thiuram	0.48
8. DM/C	0.18
9. Sulfur	0.36
Total	**39.90**

Note:
Specific gravity = 1.21; Wallace plasticity = 38; Mooney scorch time (MS 1+3) at
120°C = 13 minutes; Tensile strength = 110 kg/cm^2; Ultimate elongation = 700%;
Modulus at 300% elongation = 45 kg/cm^2; Shore hardness = 58°A.

A 5 mm thick rubber sheet is cured in an autoclave for 4 hours at 140°C and used for lining with a cold bond adhesive.

14.34 Lining for Road Tankers for 80% Phosphoric Acid or 32% Hydrochloric Acid

Ingredient	phr
1. RMA 1X	80.00
2. Reclaimed rubber	20.00
3. Oil	3.00
4. Zinc oxide	5.00
5. Stearic acid	2.50
6. SRF black	45.00
7. China clay	50.00
8. Nonox B	1.50
9. Sulfur	2.50
10. MBT	1.20
11. TMT	1.20
12. ZDC	0.10
Total	**212.00**

Note:
Cure in an autoclave for 3–4 hours at 130°C.
Specific gravity = 1.23; Wallace plasticity = 30; Mooney scorch time (MS 1+3) at 120°C = 10 minutes; Tensile strength = 140 kg/cm^2; Ultimate elongation = 500%; Modulus at 300% elongation = 65 kg/cm^2; Shore hardness = 55–60°A.

Reclaimed rubber is added as weight-for-weight substitution of total rubber to aid processing and not on the basis of rubber hydrocarbon, which is the normal practice.

15 Grooved Rubber Pads for Railways

15.1 Introduction

Rubber pads are used between wooden sleepers and rails in railway tracks. Asian and European countries have good market potential for this product.

Ingredient		Alternate I (phr)	Alternate II (phr)
1.	Smoked sheet	100.00	100.00
2.	Zinc oxide	5.00	5.00
3.	Stearic acid	2.50	2.50
4.	Elasto 710	5.00	5.00
5.	Phenyl beta-naphthylamine	1.00	1.00
6.	Paraffin wax	1.00	1.00
7.	Fast extrusion furnace black	45.00	45.00
8.	China clay	20.00	—
9.	Vulcacit CZ	0.75	0.75
10.	Tetramethylthiuram disulfide	1.25	1.25
11.	Sulfur	1.50	1.50
	Total	**183.00**	**163.00**

Note:
Cure for 10 minutes at 140°C.
Specific gravity = 1.30 (alternate I), 1.22 (alternate II); Wallace plasticity = 30, 28; Mooney scorch time (MS 1+3) at 120°C = 10 minutes, 10 minutes; Tensile strength = 150 kg/cm^2, 165 kg/cm^2; Ultimate elongation = 500%, 550%; Modulus at 300% elongation = 60 kg/cm^2, 58 kg/cm^2; Shore hardness = 60–65°A, 55–60°A

16   Paddy Dehusking Rolls

16.1   Formulation Based on Natural Rubber

Ingredient	phr
1. RMA 1X (40 Mooney)	70.00
2. Cisamer 1203	30.00
3. Zinc oxide	5.00
4. Stearic acid	1.00
5. Silica	90.00
6. Diethylene glycol	1.00
7. Coumarone-indene (CI) resin	5.00
8. Antioxidant	1.00
9. Accelerator CZ	1.50
10. Tetramethylthiuram disulfide (TMT)	0.50
11. Sulfur	3.00
12. Salicylic acid	0.50
Total	**208.50**

Note:
Cure for 30–45 minutes in a press at 150°C. Curing can also be done in an autoclave by wrapping with wet nylon cloth.

Specific gravity = 1.15; Wallace plasticity = 25; Mooney scorch time (MS 1+3) at 120°C = 10 minutes; Tensile strength = 200 kg/cm^2; Ultimate elongation = 700%; Modulus at 300% elongation = 55 kg/cm^2; Shore hardness = 80°A.

16.2   Formulation Based on Carboxylated Nitrile Rubber

Ingredient	phr
1. XNBR 221	85.00
2. Natural rubber RMA 1X	15.00
3. Sulfur	1.50
4. Antioxidant	1.00
5. Stearic acid	3.00
6. Silica filler	100.00
7. CI resin	20.00
8. Dibutyl phthalate	15.00
9. CZ	2.00
10. TMT	0.50
Total	**243.00**

Note:
Cure for 30 minutes at 150°C. Curing can also be done in an autoclave by wrapping with wet nylon cloth.
Wallace plasticity = 48; Mooney scorch time (MS 1+3) at 120°C = 10 minutes;
Tensile strength = 140 kg/cm^2; Ultimate elongation = 490%; Modulus at 100% elongation = 36 kg/cm^2; Shore hardness = 80°A.

17   Footwear Rubber Components

17.1   Solid Rubber Soling for Footwear

Ingredient	phr
1. SS RMA 1X	70.00
2. Butakon S 8551	30.00
3. Sulfur	2.50
4. Vulcafor F	1.50
5. Zinc oxide	5.00
6. Stearic acid	1.00
7. Fortafil A.70	50.00
8. Triethanolamine	1.00
9. Coumarone-indene (CI) resin	3.00
10. Nonox EXN	1.00
11. Vulcafor fast brown color	1.00
Total	**166.00**

Note:
Cure for 30 minutes at 150°C.
Wallace plasticity = 45; Mooney scorch time (MS 1+3) at 120°C = 10 minutes;
Tensile strength = 205 kg/cm^2; Ultimate elongation = 500%; Modulus at 300%
elongation = 70 kg/cm^2; Shore hardness = 80°A.

17.2   Black Heel for Footwear

Ingredient	phr
1. SS RMA 4	35.00
2. Whole tire reclaim	130.00
3. Pine tar	6.00
4. Stearic acid	3.00
5. Zinc oxide	5.00
6. High abrasion furnace (HAF) black	40.00
7. China clay	15.00
8. Sulfur	3.00
9. Accelerator MBTS	1.50
10. Tetramethylthiuram disulfide (TMT)	0.10
Total	**238.60**

Note:

Cure for 10–15 minutes at 153°C.

Specific gravity = 1.30; Wallace plasticity = 35; Mooney scorch time (MS 1+3) at 120°C = 10 minutes; Tensile strength = 140 kg/cm^2; Ultimate elongation = 450%; Modulus at 300% elongation = 68 kg/cm^2; Shore hardness = 75°A.

Alter the accelerator system as required to match processing conditions. Reclaimed rubber is added on the basis of rubber hydrocarbon at 50%.

17.3   Brown Soling for Footwear

Ingredient	phr
1. Smoked sheet	100.00
2. Stearic acid	2.00
3. CI resin	10.00
4. Pine tar	1.00
5. Antioxidant	1.00
6. Zinc oxide	10.00
7. Hard clay	175.00
8. Magnesium carbonate	40.00
9. Red oxide	5.00
10. Sulfur	4.00
11. MBTS	1.00
12. TMT	0.50
Total	**349.50**

Note:
Cure for 10–15 minutes at 153°C.
Specific gravity = 1.40; Wallace plasticity = 38; Mooney scorch time (MS 1+3) at 120°C = 10 minutes; Tensile strength = 130 kg/cm^2; Ultimate elongation = 400%; Modulus at 300% elongation = 80 kg/cm^2; Shore hardness = 75–80°A.

17.4   Sponge with High Styrene Nitrile Rubber for Soling

Ingredient	phr
1. Pale crepe	80.00
2. Butakon S 8551	20.00
3. Sulfur	2.50
4. Vulcafor F	0.75
5. Zinc oxide	5.00
6. Stearic acid	8.00
7. China clay	50.00
8. Fortafil A-70	25.00
9. Triethanolamine	1.00
10. Petroleum jelly	15.00
11. Nonox EXN	1.00
12. Vulcacel BN (blowing agent)	4.00
13. Vulcamel TBN	1.20
14. Fast brown color	0.50
Total	**213.95**

Note:
Cure for 8 minutes at 150°C. Post-cure for 3 hours at 100°C.

17.5   Sponge Rubber Soling for Footwear (Natural Rubber)

Ingredient	phr
1. SS RMA 1X	100.00
2. Sulfur	2.50
3. Vulcafor F	0.75
4. Zinc oxide	5.00
5. Stearic acid	5.00
6. China clay	75.00
7. Petroleum jelly	15.00
8. Nonox EXN	1.00
9. Vulcamel TBN	1.00
10. Vulcacel BN	3.00
11. Fast brown color	1.00
Total	**209.25**

Note:
Cure for 8 minutes at 150°C. Post-cure for 3 hours at 100°C.

17.6   Rubber Strap for Microcellular/Sponge Soling

Ingredient	phr
1. RMA 4	100.00
2. Zinc oxide	5.00
3. Stearic acid	2.00
4. Activated $CaCO_3$	30.00
5. China clay	10.00
6. Rosin	5.00
7. Paraffin wax	1.50
8. Titanium dioxide	4.00
9. Nonox SP	1.00
10. Nonox B	0.50
11. Accicure F	1.00
12. TMT	0.50
13. Sulfur	2.25
Total	**162.75**

Note:
Cure for 5 minutes at 155°C.
Add color as required.
Specific gravity = 1.36; Wallace plasticity = 35; Mooney scorch time (MS 1+3) at 120°C = 12 minutes; Tensile strength = 120 kg/cm²; Ultimate elongation = 450%; Modulus at 300% elongation = 80 kg/cm²; Shore hardness = 55°A.

18 Hoses

18.1 Introduction

The chief function of a hose is to carry air, water, oil, dry cement, sand, slurry, steam, many other chemicals, acids, and materials of various kinds where the use of a rigid pipe is impractical. In one respect a hose is merely a substitute for a pipe. In another, it is a super pipe that possesses, among other valuable properties, flexibility permitting it to be bent and moved easily and quickly from one place to another. The general structure of a hose consists of an inner functional layer and an outer layer which in most cases is reinforced with a fabric, steel, or nylon cord.

18.2 Nitrile Rubber Hose Outer

Ingredient	phr
1. Butakon AC 5502	100.00
2. Sulfur	1.75
3. Vulcafor F	1.25
4. Zinc oxide	3.00
5. Stearic acid	1.00
6. Fine thermal (FT) carbon black	80.00
7. General purpose furnace (GPF) black	40.00
8. Dioctyl phthalate	15.00
9. Hexaplas CMV	20.00
10. Nonox BL	2.00
Total	**264.00**

Note:
Cure for 30 minutes at 150°C.
Wallace plasticity = 38; Mooney scorch time (MS 1+3) at 120°C = 12 minutes; Tensile strength = 110 kg/cm^2; Ultimate elongation = 375%; Modulus at 100% elongation = 60 kg/cm^2; Shore hardness = 70°A.

18.3   Nitrile Rubber Hose Inner

Ingredient	phr
1. Butakon A 3051	100.00
2. Sulfur	1.50
3. Vulcafor F	1.25
4. Zinc oxide	5.00
5. Stearic acid	1.00
6. FT carbon black	80.00
7. GPF black	60.00
8. Dioctyl phthalate	10.00
9. Hexaplas LMV	25.00
10. Nonox BL	2.00
Total	**285.75**

Note:
Cure for 30 minutes at 150°C.
Wallace plasticity = 38; Mooney scorch time (MS 1+3) at 120°C = 12 minutes;
Tensile strength = 100 kg/cm^2; Ultimate elongation = 350%; Modulus at 300%
elongation = 57 kg/cm^2; Shore hardness = 75°A.

19   Typical Ebonite Formulations .

19.1   Introduction

Ebonites are hard rubber products obtained by the prolonged vulcaniza-tion of rubber with large proportions of sulfur.

The important properties of ebonite are:

1. Toughness combined with hardness
2. Solvent and chemical resistance
3. Ease of machining
4. Glossy look
5. Excellent electrical resistance

A typical ebonite formulation is given below:

Ingredient		phr
1.	RMA 4	100.00
2.	Sulfur	42.00
3.	Vulcafor BA	1.50
4.	China clay	65.00
5.	Barytes	35.00
6.	Lime stone	20.00
7.	Vulcatac CH	3.00
8.	Mineral oil	2.00
	Total	**268.50**

Note:
Cure for 2 hours at 150°C in steps.
Specific gravity = 1.30; Shore hardness = 80°D.

Lime is added to neutralize the acidic gases evolved during vulcanization. To offset the black color of the product, high level of titanium dioxide and fast color (i.e., 50 phr and 15 phr, respectively) can be added. Synthetic rubber gives reliable colored ebonite.

19.2 Fast Curing Ebonite

Ingredient	phr
1. RMA 1X	20.00
2. Whole tire reclaim	160.00
3. Sulfur	45.00
4. DHC	1.50
5. DAU	1.50
6. China clay	65.00
7. Ebonite dust	50.00
8. Barytes	35.00
9. Lime	20.00
10. Vulcatac CH	3.00
11. Process oil	7.00
12. Graphite	15.00
Total	**423.00**

Note:

This compound is a low cost ebonite that is used for making ebonite parts.

Cure for 2 hours at 150°C in steps.

Specific gravity = 1.35; Shore hardness = 80°D.

Three formulations for table mats based on cost are given below:

Ingredient	Low Cost (phr)	Average Cost (phr)	Quality (phr)
1. SS RMA 1X	100.00	100.00	100.00
2. Sulfur	3.25	3.25	3.25
3. Vulcafor DHC	1.50	1.50	1.50
4. Zinc oxide	5.00	5.00	5.00
5. Stearic acid	1.50	1.50	1.50
6. China clay	75.00	70.00	50.00
7. Whiting	100.00	50.00	30.00
8. Mineral oil	2.00	2.00	2.00
9. Paraffin wax	1.00	1.00	1.00
10. Nonox CNS	—	0.25	1.00
11. Titanium dioxide	2.00	5.00	10.00
12. Fast color	1.50	1.00	1.00
Total	**292.75**	**240.50**	**206.25**

Note:
Cure for 10 minutes at 150°C.
Specific gravity = 1.40 (low cost, average cost, and quality); Shore hardness = 70°A
(low cost, average cost, and quality).

21 Rubber Erasers

21.1 Introduction

Rubber erasers are one of the earliest uses of rubber since its discovery. The market offers a wide range of erasers. Being malleable and elastic, rubber can be altered in its form and stretched. It is not damaged when rubbed against paper. Because rubber has an affinity for carbon, from which pencils and sometimes pens are made, it absorbs carbon when rubbed against the pencil/pen marks made on paper. It is mind-boggling to imagine the quantity of rubber erasers used in schools, colleges, offices, drawing offices, etc. How many children use them! How many adults use them! Almost everyone, everywhere, worldwide, most of the time! Can we imagine life without erasers? Can we rub out all our mistakes without spoiling the paper?

Some rubber eraser formulations are given below:

21.2 Pencil Eraser—Alternate I

Ingredient	phr
1. Pale crepe	100.00
2. Sulfur	6.00
3. Vulcafor EFA	4.00
4. Zinc oxide	8.00
5. Stearic acid	1.00
6. Light calcined magnesia (MgO)	35.00
7. Ammonium oleate	17.00
8. Rubber crumbs	125.00
9. White factice	300.00
10. Whiting	250.00
11. Nonox WSL	1.50
12. Titanium dioxide	5.00
13. Fast red color	4.00
Total	**856.50**

Note:
The rubber content in the above formulation is only about 10%.
Cure for 8–10 minutes at 150°C.

21.3   Pencil Eraser—Alternate II

Ingredient	phr
1. Natural Rubber RMA 1X	100.00
2. White factice	200.00
3. Light calcined MgO	30.00
4. Whiting	150.00
5. Barytes	150.00
6. Titanium dioxide	5.00
7. Zinc oxide	10.00
8. Stearic acid	2.00
9. Process oil	60.00
10. Hydrated lime	25.00
11. Nonox SP antioxidant	1.50
12. Accicure F	3.50
13. Diethylene glycol	1.00
14. Sulfur	8.00
Total	**746.00**

Note:
Add color as required.
Cure for 8 minutes at 150°C.

21.4   Ink Eraser—Alternate I

Ingredient	phr
1. Pale crepe	100.00
2. Sulfur	6.00
3. Vulcafor F	1.00
4. Vulcafor TMT	0.60
5. Light calcined MgO	10.00
6. Stearic acid	1.00
7. White factice	50.00
8. Lime	60.00
9. Mineral oil	9.00
10. Nonox EXN	1.00
11. Glass powder (silica)	200.00
12. Titanium dioxide	2.00
13. Fast color	2.00
Total	**442.60**

Note:
Glass powder improves abrasion.
Cure for 8–10 minutes at 150°C.

21.5   Ink Eraser—Alternate II

Ingredient		phr
1.	Natural rubber RMA 4	100.00
2.	White factice	150.00
3.	Light calcined MgO	25.00
4.	Whiting	150.00
5.	Barytes	150.00
6.	Titanium dioxide	5.00
7.	Zinc oxide	10.00
8.	Stearic acid	2.00
9.	Process oil	10.00
10.	Hydrated lime	20.00
11.	Nonox SP	1.00
12.	Accicure F	3.50
13.	Diethylene glycol	1.00
14.	Sulfur	8.00
15.	Carborundum powder	50.00
	Total	**685.50**

Note:
Carborundum powder improves abrasion.
Cure for 8–10 minutes at 150°C.

21.6   Eraser for Typewriter

Ingredient	phr
1. SS RMA 4	100.00
2. White factice	200.00
3. Light calcined MgO	30.00
4. Whiting	125.00
5. Barytes	125.00
6. Titanium dioxide	5.00
7. Zinc oxide	10.00
8. Stearic acid	2.00
9. Process oil	60.00
10. Hydrated lime	25.00
11. Nonox SP	1.00
12. Accicure	3.50
13. Diethylene glycol	1.00
14. Sulfur	8.00
15. Carborundum powder	100.00
Total	**795.50**

Note:
Cure for 8–10 minutes at 150°C.

Typewriter marks require higher abrasion and so the proportion of carborundum powder is higher; the proportion of factice is also higher to prevent damage to paper.

22 Natural Rubber (NR) Study Formulations—Factory Trials

Ingredient		NR1 (phr)	NR2 (phr)	NR3 (phr)	NR4 (phr)
1.	RMA 1X	100.00	100.00	100.00	100.00
2.	Zinc oxide	5.00	5.00	5.00	5.00
3.	Nonox HFN	1.00	1.00	1.00	1.00
4.	Stearic acid	1.25	1.50	3.00	1.50
5.	Channel black	—	—	45.00	—
6.	Semi-reinforcing furnace black	40.00	30.00	—	37.50
7.	China clay	—	75.00	—	25.00
8.	MBTS	1.00	1.00	1.00	1.00
9.	Sulfur	2.50	3.00	2.50	3.00
10.	Vulcafor BSO	0.25	0.25	—	0.25
11.	Dutrex R	1.00	2.00	5.00	2.00
	Total	**152.00**	**218.75**	**162.50**	**176.25**

Note:
Cure time = 30 minutes at 145°C (NR1 and NR2), 40 minutes at 150°C (NR3), 20 minutes at 145°C (NR4).

Physical Properties:

		NR1	NR2	NR3	NR4
1.	Tensile strength (psi)	2700	1746	3287	2370
2.	Modulus at 100% elongation (psi)	270	411	280	386
3.	Modulus at 300% elongation (psi)	1200	1158	1071	1251
4.	Ultimate elongation (%)	550	420	590	480
5.	Shore hardness (°A)	55	66	61	62
6.	Rebound resilience (%) (DIN standard)	50	43	29	48

Note:
Compare the physical properties with changes in doses and types of fillers, accelerators, and process oils. These compounds can be used for making molded products.

23 White Rubber Tiles

Ingredient	phr
1. Pale crepe	100.00
2. Stearic acid	1.50
3. Paraffin wax	2.00
4. Aromatic plasticizer	1.00
5. Zinc oxide	10.00
6. Soft clay	275.00
7. Whiting	50.00
8. Titanium dioxide	25.00
9. Sulfur	4.00
10. MBTS	1.00
11. Tetramethylthiuram disulfide (TMT)	0.10
Total	**469.60**

Note:
Cure for 10–15 minutes at 153°C.
Specific gravity = 1.50; Wallace plasticity = 35; Mooney scorch time (MS 1+3)
at 120°C = 10 minutes; Tensile strength = 110 kg/cm^2; Ultimate elongation = 380%;
Modulus at 300% elongation = 80 kg/cm^2; Shore hardness = 80°A.

24 Factory Trials of Neoprene Moldables

24.1 Introduction

The following factory trials of neoprene compounds with changes in the types of fillers and plasticizers and their loadings along with changes in the dosage levels of the accelerator system give an interesting comparative study of physical properties. The formulations for these compounds are given below to enable progressive compounding development work. These compounds can be suitable for molding applications.

Ingredients	1 (kg)	2 (kg)	3 (kg)	4 (kg)	5 (kg)	6 (kg)	7 (kg)
1. Neoprene GRT	2.000	2.000	2.000	2.000	—	—	—
2. Neoprene WRT	—	—	—	—	2.0000	2.000	2.000
3. RMA 1X	1.000	1.000	—	0.500	0.5000	0.300	0.300
4. Reclaimed rubber	—	—	1.000	0.500	0.5000	—	—
5. Nonox BL	0.030	0.020	0.020	0.020	0.0200	0.020	0.020
6. Nonox D	—	—	—	0.020	0.0200	—	—
7. Flectol H	—	0.020	0.020	—	—	0.020	0.010
8. Stearic acid	0.030	0.040	0.030	0.030	0.0250	0.025	0.013
9. Light calcined magnesia	0.040	0.040	0.040	0.080	0.0800	0.080	0.080
10. Zinc oxide	0.200	0.200	0.200	0.200	0.2000	0.200	0.120
11. Dutrex RT	0.250	0.375	0.250	0.375	0.3000	0.400	0.400
12. Dutrex R	—	—	0.200	0.125	0.2000	0.150	0.100
13. Hexaplas PPL	0.100	0.150	0.150	0.150	0.1500	0.100	0.100
14. Vulcatac CH	—	—	0.050	—	—	—	—
15. Phil black G	0.750	0.200	0.350	0.400	—	—	—
16. Phil black A	0.250	0.200	0.200	0.100	0.1000	—	—
17. Kosmos 20	—	0.200	0.200	0.100	0.9000	0.750	0.850
18. China clay	0.750	0.750	0.750	1.000	1.0000	0.400	0.750
19. Whiting	—	0.250	0.250	0.250	0.2500	0.300	0.250
20. Sulfur	0.045	0.045	0.045	0.035	0.0375	0.018	0.030
21. Tetramethylthiuram disulfide (TMT)	0.020	0.020	0.020	0.015	0.0235	0.015	0.017
Total	5.465	5.510	5.775	5.900	6.3060	4.778	5.040

Note:
Cure time = 20 minutes at 150°C.

Physical Properties:

		1	2	3	4	5	6	7
1.	Tensile strength (psi)	2340	1990	1170	1297	1160	1280	1173
2.	Modulus at 300% elongation (psi)	1380	860	930	871	810	700	—
3.	Modulus at 100% elongation (psi)	450	295	380	335	285	255	—
4.	Ultimate elongation (%)	480	550	356	397	436	484	450
5.	Tear resistance (lb/in.)	205	165	284	273	270	425	190
6.	Shore hardness (°A)	63	58	60	60	58	53	55
7.	Rebound resilience (%)	39	39	22	32	25	36	32

Note:
These results are obtained from conventional testing equipments and they provide values for a good comparative study of the formulations. It will be easier to compare the results with the formulations if the dosages are converted to phr levels.

25 Proofing Compounds for Clothing and Inflatables

25.1 Introduction

Rubber coated fabrics are widely used in clothing and water proofing applications. Present interest in these materials includes inflatable products ranging from air–sea rescue equipment to vehicle-recovery bags. Natural rubber is used to a large extent for this application. A more recent application of proofed fabrics includes inflatables for lifting vehicles and aeroplanes.

25.2 Frictioning Compound

Ingredient	phr
1. Natural rubber	100.00
2. Brown or white factice (as required)	45.00
3. Calcium carbonate	100.00
4. Barium sulfate	80.00
5. Zinc oxide	10.00
6. MBTS	1.80
7. Sulfur	3.60
Total	**340.40**

25.3 Topping Compound

Ingredient	phr
1. Natural rubber	100.00
2. Zinc oxide	5.00
3. Brown or white factice (as required)	50.00
4. Whiting	100.00
5. China clay	50.00
6. Antioxidant	1.00
7. Accelerator F	0.75
8. Sulfur	3.00
Total	**309.75**

Note:
The fabric is first fractioned and then topped.

26 Wear Resistant Rubber for the Mining Industry

26.1 Introduction

Mining and ore handling industries encounter a wide variety of minerals and fluids containing slurries with abrasive solids of ore particles. Equipments that are exposed to the abrasive environment are prone to extreme wear. Many years of experience has proved that the use of rubber is the solution for applications where high resistance to abrasive wear is required. The reason for the high wear resistance of rubber is its flexibility and elasticity. The impinging particles of ore cause deformation and wear even in the hardest metal surfaces. In the case of rubber it yields and absorbs the impacts, regaining its original shape immediately, and as a result there is very little wear. Natural rubbers, such as pure gum, with a hardness of 40°A have shown very good endurance to sliding abrasion in sand and slurry handling pipes and pumps, where particles have fine size and no grit is present. Where resistance to tearing is required, rubber must be compounded with carbon black to give a tough composition of compounds with a hardness of 60–65°A. Such linings are used in chutes, hoppers, and launders. In ore handling industries it has been found that properly designed rubber compounds can outlast metal surfaces by 10 to 1 in many cases.

26.2 Typical Slurry Handling Compound—40°A

Ingredient	phr
1. RMA 1X	100.00
2. Semi-reinforcing furnace (SRF) black	40.00
3. Zinc oxide	5.00
4. Stearic acid	3.00
5. Mercaptobenzothiazole (MBT)	1.00
6. Elasto 710	3.00
7. Sulfur	2.75
8. Phenyl beta-naphthylamine (PBNA)	1.00
Total	**155.75**

Note:
Cure for 3 hours at 140°C in an autoclave.
Specific gravity = 1.20; Wallace plasticity = 32; Mooney scorch time (MS 1+3) at 120°C = 8 minutes; Tensile strength = 200 kg/cm^2; Ultimate elongation = 650%; Modulus at 300% elongation = 80 kg/cm^2; Shore hardness = 40°A.

26.3 Typical Chute and Launder Lining Compound—60°A

Ingredient	phr
1. RMA 1X	100.00
2. Zinc oxide	5.00
3. Stearic acid	3.00
4. Fast extrusion furnace black	20.00
5. SRF black	40.00
6. Precipitated CaCO$_3$	15.00
7. Elasto 710	4.00
8. MBT	0.50
9. Vulcatard A	0.50
10. Sulfur	2.75
11. PBNA	1.00
Total	**191.75**

Note:
Cure for 3 hours at 140°C in an autoclave.
Specific gravity = 1.25; Wallace plasticity = 35; Mooney scorch time (MS 1+3) at 120°C = 8 minutes; Tensile strength = 210 kg/cm^2; Ultimate elongation = 600%; Modulus at 300% elongation = 90 kg/cm^2; Shore hardness = 60°A.

27 Neoprene Molded Corks

27.1 Introduction

These corks are used in chlor-alkali industries as consumables to plug pipe and tank outlets during cleaning operations.

Ingredient	phr
1. Neoprene GRT	100.00
2. RMA 1X	15.00
3. Flectol H	2.00
4. Stearic acid	1.50
5. Magnesium oxide	4.00
6. Zinc oxide	10.00
7. Aromatic process oil	25.00
8. Semi-reinforcing furnace black	10.00
9. General purpose furnace black	25.00
10. China clay	25.00
11. Whiting	15.00
12. Sulfur	1.00
13. Tetramethylthiuram disulfide	0.75
Total	**234.25**

Note:
Cure for 10 minutes at 150°C.
Specific gravity = 1.50; Wallace plasticity = 40; Mooney scorch time (MS 1+3) at 120°C = 10 minutes; Tensile strength = 148 kg/cm^2; Ultimate elongation = 750%; Modulus at 300% elongation = 50 kg/cm^2; Shore hardness = 60°A.

Natural rubber is added for easy processing and for its resilience.

28 Low-Cost Chemical Resistant Canvass Reinforced Neoprene Rubber Sheets

Ingredient	phr
1. Neoprene WB	50.00
2. Styrene-butadiene rubber (general purpose grade)	50.00
3. Phenyl beta-naphthylamine	1.00
4. Stearic acid	1.50
5. Magnesium oxide	1.00
6. Zinc oxide	5.00
7. Semi-reinforcing furnace black	60.00
8. China clay (fine)	70.00
9. Aromatic process oil	30.00
10. Sulfur	0.75
11. Vulcacit CZ	0.75
12. Tetramethylthiuram disulfide	0.75
Total	**270.75**

Note:
Cure for 2 hours at 150°C in an autoclave.
Specific gravity = 1.50; Wallace plasticity = 38; Mooney scorch time (MS 1+3) at 120°C = 12 minutes; Tensile strength = 110 kg/cm^2; Ultimate elongation = 500%; Modulus at 300% elongation = 50 kg/cm^2; Shore hardness = 65°A.

Ingredient	phr
1. SBR 1712	25.00
2. Whole tire reclaim	160.00
3. Pine tar	15.00
4. Sulfur	45.00
5. Accicure F	3.00
6. Lime	25.00
7. China clay	50.00
8. Ebonite dust	70.00
9. Paraffin wax	2.00
10. Magnesium oxide	5.00
Total	**400.00**

Note:
Cure for 25 minutes at 160°C and post-cure in an autoclave.
Specific gravity = 1.40; Shore hardness = 85°D.

30   Neoprene Washer for Water Taps

Ingredient	phr
1. Neoprene WRT	100.00
2. Zinc oxide	10.00
3. Nonox BL	1.00
4. Stearic acid	0.50
5. General purpose furnace black	15.00
6. High abrasion furnace black	15.00
7. Light calcined MgO	4.00
8. China clay	75.00
9. Whiting	25.00
10. Tetramethylthiuram disulfide	0.10
11. Sulfur	0.75
12. Dutrex R	8.00
13. Paraffin wax	1.00
Total	**255.35**

Note:
Cure for 15 minutes at 150°C.
Specific gravity = 1.55; Wallace plasticity = 30; Mooney scorch time (MS 1+3) at 120°C = 10 minutes; Tensile strength = 140 kg/cm^2; Ultimate elongation = 650%; Modulus at 300% elongation = 70 kg/cm^2; Shore hardness = 75°A.

31   Neoprene Inner Layer for Isocyanate Bonded Components

Ingredient	phr
1. Neoprene WRT	100.00
2. Zinc oxide	5.00
3. Nonox BL	1.50
4. Stearic acid	2.50
5. General purpose furnace black	30.00
6. Light calcined MgO	4.00
7. Tetramethylthiuram disulfide	0.50
8. Dutrex R	3.00
Total	**146.50**

Note:
Cure for 20 minutes at 153°C.
Specific gravity = 1.38; Wallace plasticity = 35; Mooney scorch time (MS 1+3) at 120°C = 20 minutes; Tensile strength = 125 kg/cm^2; Ultimate elongation = 430%; Modulus at 300% elongation = 80 kg/cm^2; Shore hardness = 66°A.

This compound has no sulfur. It is used as an inner layer in isocyanate or ebonite bonded products as a cushion. The cushioning effect is helpful in metal–rubber bonded shock mounts. About 1 phr of sulfur may be required if the material surface is made of brass.

32 Rubber Bonded Anvil for the Electronics Industry

	Ingredient	phr
1.	Neoprene WRT	100.00
2.	Zinc oxide	10.00
3.	Nonox HFN	2.00
4.	Stearic acid	0.50
5.	Semi-reinforcing furnace (SRF) black	75.00
6.	High abrasion furnace black	25.00
7.	Light calcined MgO	4.00
8.	Tetramethylthiuram disulfide	1.00
9.	Sulfur	0.50
10.	Dutrex R	10.00
	Total	**228.00**

Note:
Cure for 30 minutes at 153°C.
Specific gravity = 1.48; Wallace plasticity = 40; Mooney scorch time (MS 1+3) at 120°C = 15 minutes; Tensile strength = 215 kg/cm^2; Ultimate elongation = 330%; Modulus at 300% elongation = 180 kg/cm^2; Shore hardness = 80°A.

Change SRF black to general purpose furnace black for increased hardness.

33　Solid Tires for Forklift Trucks

33.1　Introduction

Low rolling resistance is required for solid tires used in material handling equipment such as forklift trucks. Care should to be taken in the choice of reinforcing fillers, accelerator system, and plasticizer levels. The following formula is suggested for a tightly cured solid tire.

Ingredient	phr
1.　RMA 1X	100.00
2.　High abrasion furnace black	50.00
3.　Paraffin wax	1.00
4.　Aromatic oil	5.00
5.　Nonox BL	1.00
6.　Nonox HFN	1.00
7.　Zinc oxide	4.00
8.　Zinc stearate	7.00
9.　MBTS	1.50
10.　Sulfur	1.50
11.　Retarding agent	0.10
Total	**172.10**

Note:
Cure for 30 minutes at 150°C.
Specific gravity = 1.23; Wallace plasticity = 35; Mooney scorch time (MS 1+3) at 120°C = 9 minutes; Tensile strength = 250 kg/cm^2; Ultimate elongation = 495%; Modulus at 300% elongation = 140 kg/cm^2; Shore hardness = 65°A.

For improved heat resistance the accelerator system has to be changed to MS or tetramethylthiuram disulfide. This also gives reversion resistance resulting from long vulcanization times in the case of thicker tires.

34 Pharmaceutical Bottle Closures

34.1 Introduction

Compounds for pharmaceutical bottle closures should have low extractable content and should withstand repeated sterilization in super-heated steam/water. These compounds should be tested for physiological effects and conformity to health regulations before manufacture.

Ingredient	A (phr)	B (phr)	C (phr)
1. Polysar Butyl 301	100.00	100.00	—
2. Polysar Butyl 402	—	—	100.00
3. Zinc oxide	3.00	3.00	5.00
4. Barytes	50.00	50.00	50.00
5. Titanium dioxide	15.00	15.00	15.00
6. Anhydrous silica	30.00	30.00	30.00
7. Paraffin wax	2.00	2.00	2.00
8. Polyethylene	—	—	2.00
9. Vulcacit P	1.50	—	0.70
10. Vulcacit PN	0.20	—	0.50
11. Zinc dibutyldithiocarbamate	—	1.60	0.60
12. Benzothiazyl disulfide	—	0.20	—
13. Mercaptobenzothiazole	—	—	0.20
14. Sulfur	1.20	1.20	1.00
Total	**202.90**	**203.00**	**207.00**

Note:
Cure for 20 minutes at 160°C (for A, B, and C).
Specific gravity = 1.10; Mooney scorch time (MS 1+3) at 120°C = 12 minutes; Tensile strength = 60 kg/cm^2; Ultimate elongation = 600%; Modulus at 300% elongation = 12 kg/cm^2; Shore hardness = 45°A.

APPENDICES

Appendix 1: Scorching of Rubber—A Study Formula

Scorching is the premature vulcanization of a product at an undesirable stage of processing such as mixing, molding, calendering, extrusion, and other forming operations.

The compound formula given below will scorch even at a temperature of 100°C and become unusable for further processing such as mixing, calendaring, extrusion, and molding.

Ingredient	phr
1. SS RMA 1X	100.00
2. Sulfur	3.00
3. Vulcafor MBT	2.00
4. Diphenyl guanidine	2.00
5. Zinc oxide	5.00
6. Stearic acid	2.00
7. Barytes	75.00
8. Petroleum jelly or any other plasticizer	5.00
9. Nonox EXN	1.00
Total	**195.00**

This experimental formulation is given to alert the rubber technologist to avoid a wrong dosage of accelerator being used. However, scorching occurs when processing temperatures are high, due to the accelerating effects of certain fillers and other ingredients such as antioxidants, process aids, and high dosage of inorganic oxides.

Appendix 2: Specific Gravity and Volume Cost

Rubber goods are generally sold by the piece. In other words, they are sold based on volume and not weight. However, certain cheap products such as low cost sheets are sold by weight. Therefore, it is important that the rubber technologist knows the specific gravity of the product and also those of the ingredients used. The lower the specific gravity of an ingredient, the smaller the weight that is required to fill a definite volume. Practically all rubber ingredients are sold by weight, and therefore the rubber technologist should always be on the alert to arrive at a low weight cost, together with low specific gravity, resulting in a low volume cost. The information regarding the specific gravity of the ingredients can be obtained from the seller or manufacturer.

The specific gravity of a rubber compound with the following formulation can be calculated as follows:

Ingredient	phr	Specific Gravity	Volume
1. SS RMA 1X	100.00	0.92	108.60
2. HFN antioxidant	1.00	1.22	0.82
3. Zinc oxide	5.00	5.55	0.90
4. MBTS	1.00	1.54	0.65
5. Stearic acid	3.00	0.85	3.53
6. Dutrex R	5.00	1.01	4.96
7. Sulfur	2.50	2.10	1.19
8. High abrasion furnace carbon black	45.00	1.80	25.00
Total	**162.50**		**145.65**

Specific gravity of the compound = Weight/Volume = 162.5/145.65 = 1.115.

Appendix 3: Equivalent Chemical Names for Trade Names

Trade Name	Chemical Description	Supplier
Accelerator MC (Nonox NSN)	Phenol-aldehyde-amine condensate	
Accelerator ZMBT	Zinc salt of mercaptobenzothiazole	Dupont
Accicure F	Benzothiazyl disulfide and diphenyl guanidine	ACCI
Accicure HBS	N-Cyclohexyl-2-benzothiazyl sulfenamide	ACCI
Accinox DN	Antioxidant	ACCI
Accinox HFN	Blend of arylamines	ACCI
Accinox TQ	Antioxidant	ACCI
Accitard A	N-Nitrosodiphenylamine (retarder)	ACCI
Ancoplas ER	Mixture of sulfonated petroleum products (peptizer)	Anchor
Antioxidant NBC	Nickel dibutyl dithiocarbamate	Dupont/Uniroyal
Antioxidant NBC	Nickel dibutyl dithiocarbamate	Dupont/Uniroyal
Antioxidant PBNA	Phenyl beta-naphthylamine	Anchor
Bayprene 110	Chloroprene equivalent to Dupont's Neoprene WRT	Bayer
Butakon A 3051	Medium oil resistant nitrile	ICI
Butakon AC 5502	Blend of acrylonitrile and polyvinyl chloride	ICI
Butakon S 8551	Styrene-butadiene copolymer with 85% styrene-reinforcing resin	ICI
Butyl 268	Inner tube grade butyl rubber	Exxon
Celite PF3	Kaoline or fine China clay	
Cisamer 1203	cis-Polybutadiene rubber	
DAU	Mixture of dibenzothiazyl and thiuram disulfide	ICI
Desmodur-R	20% solution of triphenylmethane tri-isocyanate in methylene chloride	Bayer
DHC	Blend of mercaptobenzothiazole and dithiocarbamate	ICI
Diak No. 1	Hexamethylenediamine carbonate	Dupont
Dicup 40	Dicumyl peroxide	Hercules
Dutrex R	Aromatic process oil	Shell
Dutrex RT	Aromatic oil of high molecular weight	Shell Chemicals
Elasto 541	Aromatic process oil	Exxon
Elasto 641	Naphthenic petroleum oil	Exxon
Elasto 710	Aromatic process oil	Exxon

Trade Name	Chemical Description	Supplier
EPCAR 346	Low Mooney EPDM	BF Goodrich
Epikoté resin 828	Epoxy resin	
Felxon 840	Paraffinic petroleum oil	Exxon
Flectol H	Polymerized dihydrotrimethyl quinoline (antioxidant)	Monsanto
Flexon 310	Naphthenic process oil	Exxon
Fortafil A-70	Precipitated aluminum silicate	ICI
Hexaplas LMV	Modified polypropylene adipate of medium viscosity	ICI
Hexaplas PPL	Modified polypropylene adipate	ICI
Hisil	Precipitated hydrated silica	Horwick Chem
Hycar 1001	High acrylonitrile rubber	BF Goodrich
Hycar 1042	Medium-high acrylonitrile rubber	BF Goodrich
Hycar 2202	Brominated butyl rubber	BF Goodrich
Hypalon	Chlorosulfonated polyethylene	Dupont
Koresin	Alkylphenol-acetylene resin	Akron Chem
Kosmos 20	Semi-reinforcing furnace (SRF) black	United
Krynac 801	High acrylonitrile rubber (nonstaining)	Polymer
Krynac 803	Plasticized nitrile rubber (nonstaining)	Polymer
LDP	Low density polyethylene	
MBTS	Dibenzothiazyl disulfide	ICI
NA22	2-Mercaptoimidazoline	Dupont
Neoprene AC	Chloroprene rubber (adhesive grade)	Dupont
Neoprene GRT	Sulfur modified chloroprene	Dupont
Neoprene KNR	Adhesive grade neoprene	Dupont
Neoprene WB	Smooth processing chloroprene	Dupont
Neoprene WHV	High Mooney chloroprene	Dupont
Neoprene WRT	Chloroprene (dry rubber)	Dupont
Neoprene WX	Chloroprene rubber (nonstaining)	Dupont
Nono X HSL	Antioxidant	
Nonox B	Acetone-diphenylamine condensation product (powder form)	ICI
Nonox BL	Acetone-diphenylamine condensation product (liquid form)	ICI
Nonox CNS	A blend of mercaptobenzimidazole and Nonox WSL (phenol condensation product) copper inhibitor antioxidant	ICI
Nonox D	Phenyl beta-naphthylamine	ICI
Nonox EXN	Phenol condensation product	ICI
Nonox HFN	Blend of arylamines	ICI
Nonox NSN	Phenol-aldehyde-amine condensate	ICI
Nonox SP	Styrenated phenol	ICI
Nordel	Ethylene-propylene diene monomer (EPDM)	Dupont
Octamine	Reaction product of diphenylamine and di-isobutylene	Uniroyal

Trade Name	Chemical Description	Supplier
P33 Black	Fine thermal (FT) black	RT Vanderbilt
Paraceril-B	Medium-low acrylonitrile rubber	Uniroyal
Paracril Ozo	70% NBR + 30% PVC blend	Uniroyal
Phil black "A" (FEF)	Fast extrusion furnace (FEF) black	Philips
Phil black "O" (HAF)	High abrasion furnace (HAF) black	Philips
Plasticator FH	Aromatic polyether	Mobay Chem
Polysar 300	Butyl rubber	Polysar
Polysar 301	Butyl rubber	Polysar
Polysar Krynac NS	Nonstaining acrylonitrile rubber	Polysar
Robac 22	Ethylene thiourea	Robinson
Brothers		
Santoflex AW	A condensation product of acetone and *p*-phenetidine (ethoxy-trimethyl-dihydroquinoline) antioxidant	Monsanto
SBR 1712	Oil extended styrene-butadiene rubber	
Synaprene 1502	Styrene-butadiene general purpose rubber	Synthetic and Chemicals
Tetrone A	Dipentamethylene thiuram hexasulfide	Dupont
Thikol St	Polysulfide rubber	Thiokol Chem
Thiokol FA	Nonstaining polysulfide rubber	Thiokol Chem
Thiuram E	Tetraethylthiuram disulfide	Dupont
Tipure R610	Titanium dioxide	
TMT	Tetramethylthiuram disulfide	Monsanto
Triethanolamine	Activator	Union carbon
Viton A-HV	Copolymers of vinylidene fluoride and hexafluoropropylene	Dupont
Vulcacel BN	Dinitrosopentamethylenetetramine ina base (blowing agent)	ICI
Vulcacit CBS	Cyclohexyl benzothiazole sulfenamide	ICI
Vulcacit CZ	Benzothiazyl-2-cyclohexyl sulfenamide	Bayer
Vulcacit DM/C	Dibenzothiazyl disulfide	Bayer
Vulcacit DOTG	Di-*ortho*-tolyl guanidine	Bayer
Vulcacit DPG	Diphenyl guanidine	Bayer
Vulcacit F	Mixture of dibenzothiazyl disulfate and basic accelerator	Bayer
Vulcacit LDA	Zinc-N-diethyl dithiocarbamate	Bayer
Vulcacit MS	Tetramethylthiuram monosulfide	Bayer
Vulcacit NPV	2-Mercaptoimidazoline (ethylene thiourea)	Bayer
Vulcacit P	Pentamethylene-ammonium-N-pentamethylene-dithiocarbamate	Bayer
Vulcacit PN	Zinc-N-ethyl phenyl dithiocarbamate	Bayer
Vulcacit ZDC	Zinc dimethyl dithiocarbamate	Bayer
Vulcafor BA	Butraldehyde-aniline condensate	ICI
Vulcafor BSO	Benzothiazyl sulfenamide	ICI

Trade Name	Chemical Description	Supplier
Vulcafor EFA	Condensate of formaldehyde, ammonia, and ethyl chloride (aldehyde-amine type accelerator)	ICI
Vulcafor HBS	N-Cyclohexyl-2 benzothiazyl sulfenamide	ICI
Vulcafor MBT	Mercaptobenzothiazole	ICI
Vulcafor MS	Tetramethylthiuram monosulfide	ICI
Vulcafor TET	Tetraethylthiuram disulfide	ICI
Vulcait CBS	Cyclohexyl benzothiazyl sulfenamide	ICI
Vulcalent A	N-Nitrosodiphenylamine (retarder)	Bayer
Vulcamel TBN	Tri-beta naphthol in inert wax (peptizer)	ICI
Vulcasil S	Reinforcing precipitated silica	Bayer
Vulcatac CH	Rosin in aromatic petroleum oil	ICI
Vulcatard A	Nitrosodiphenylamine	ICI
XNBR 221	Carboxylated nitrile rubbers	

Appendix 4: Useful Conversion Tables

A4.1 Tensile Strength and Modulus Conversions (Force/Area)

kg/cm^2 $\Rightarrow$ MPa		psi $\Leftarrow$ MPa $\Rightarrow$ kg/cm^2			psi $\Rightarrow$ MPa	
0.5	0.05	14.5	0.1	1.0	1	0.01
1	0.10	29.0	0.2	2.0	5	0.03
2	0.20	43.5	0.3	3.1	10	0.07
3	0.30	58.0	0.4	4.1	15	0.10
4	0.39	72.5	0.5	5.1	20	0.14
5	0.49	87.0	0.6	6.1	25	0.17
6	0.59	101.5	0.7	7.1	30	0.21
7	0.69	116.0	0.8	8.2	35	0.24
8	0.79	130.5	0.9	9.2	40	0.28
9	0.88	145.0	1.0	10.2	45	0.31
10	0.98	159.5	1.1	11.2	50	0.34
11	1.08	174.0	1.2	12.2	55	0.38
12	1.18	188.5	1.3	13.3	60	0.41
13	1.28	203.0	1.4	14.3	65	0.45
14	1.37	217.6	1.5	15.3	70	0.48
15	1.47	232.1	1.6	16.3	75	0.52
16	1.57	246.6	1.7	17.3	80	0.55
17	1.67	261.1	1.8	18.4	85	0.59
18	1.77	275.6	1.9	19.4	90	0.62
19	1.86	290.1	2.0	20.4	95	0.66
20	1.96	362.6	2.5	25.5	100	0.69
25	2.45	435.1	3.0	30.6	150	1.03
30	2.94	507.6	3.5	35.7	200	1.38
35	3.43	580.1	4.0	40.8	250	1.72
40	3.92	652.7	4.5	45.9	300	2.07
45	4.42	725.2	5.0	51.0	350	2.41
50	4.91	797.7	5.5	56.1	400	2.76
55	5.40	870.2	6.0	61.2	450	3.10
60	5.89	942.7	6.5	66.3	500	3.45
65	6.38	1015.7	7.0	71.4	550	3.79
70	6.87	1087.8	7.5	76.5	600	4.14
75	7.36	1160.3	8.0	81.6	650	4.48

kg/cm² ⇒ MPa		psi ⇐ MPa ⇒ kg/cm²			psi ⇒ MPa	
80	7.85	1232.8	8.5	86.7	700	4.83
85	8.34	1305.3	9.0	91.7	750	5.17
90	8.83	1377.8	9.5	96.8	800	5.52
95	9.32	1450.3	10	101.9	850	5.86
100	9.81	2175.5	15	152.9	900	6.21
200	19.62	2900.7	20	203.9	1000	6.90
300	29.43	4351.0	30	305.8	2000	13.79
400	39.24	5801.3	40	407.8	3000	20.69

kg, kilogram; psi, pounds per square inch; MPa, megapascal; cm, centimeter.

A4.1.1 Tear Strength Conversions (Force/Width)

kg/cm ⇒ kN/m		kg/cm ⇐ kN/m ⇒ lb/in.			lb/in. ⇒ kN/m	
0.1	0.10	0.10	0.1	0.6	1	0.18
0.5	0.49	0.20	0.2	1.1	2	0.35
1.0	0.98	0.31	0.3	1.7	3	0.53
1.5	1.47	0.41	0.4	2.3	4	0.70
2.0	1.96	0.51	0.5	2.9	5	0.88
2.5	2.45	0.61	0.6	3.4	6	1.05
3.0	2.94	0.71	0.7	4.0	7	1.23
3.5	3.43	0.82	0.8	4.6	8	1.40
4.0	3.92	0.92	0.9	5.1	9	1.58
4.5	4.41	1.02	1.0	5.7	10	1.76
5.0	4.91	1.53	1.5	8.5	11	1.93
5.5	5.40	2.04	2.0	11.4	12	2.11
6.0	5.89	2.55	2.5	14.2	13	2.28
6.5	6.38	3.06	3.0	17.1	14	2.46
7.0	6.87	3.57	3.5	19.9	15	2.63
7.5	7.36	4.08	4.0	22.8	16	2.81
8.0	7.85	4.59	4.5	25.6	17	2.98
8.5	8.34	5.10	5.0	28.5	18	3.16
9.0	8.83	5.61	5.5	31.3	19	3.34
9.5	9.32	6.12	6.0	34.2	20	3.51
10	9.81	6.63	6.5	37.0	25	4.39
15	14.72	7.14	7.0	39.9	30	5.27
20	19.62	7.65	7.5	42.7	35	6.14
25	24.53	8.16	8.0	45.6	40	7.02
30	29.43	8.67	8.5	48.4	45	7.90
35	34.34	9.18	9.0	51.3	50	8.78
40	39.24	9.69	9.5	54.1	60	10.53
45	44.15	10.20	10	57.0	70	12.29

kg/cm ⇒	kN/m	kg/cm ⇐	kN/m ⇒	lb/in.	lb/in. ⇒	kN/m
50	49.05	15.30	15	85.4	80	14.04
60	58.86	20.40	20	113.9	90	15.80
70	68.67	25.50	25	142.4	100	17.56
80	78.48	30.60	30	170.9	200	35.11
90	88.29	35.70	35	199.4	300	52.67
100	98.10	40.80	40	227.8	400	70.22
200	196.20	45.90	45	256.3	500	87.78
300	294.30	51.00	50	284.8	600	105.33
400	392.40	61.20	60	341.8	700	122.89
500	490.50	71.40	70	398.7	800	140.44
600	588.60	81.60	80	455.7	900	158.00
700	686.70	91.80	90	512.6	1000	175.55

kg, kilogram; m, meter; N, newton; lb, pound; cm, centimeter; kN, kilonewton.

A4.2   Temperature Conversions

A4.2.1   Celsius to Fahrenheit

°C	°F	°C	°F
−80	−112.0	10	50.0
−70	−94.0	20	68.0
−60	−76.0	30	86.0
−50	−58.0	40	104.0
−40	−40.0	50	122.0
−30	−22.0	60	140.0
−20	−4.0	70	158.0
−10	14.0	80	176.0
−9	15.8	90	194.0
−8	17.6	100	212.0
−7	19.4	110	230.0
−6	21.2	120	248.0
−5	23.0	130	266.0
−4	24.8	140	284.0
−3	26.6	150	302.0
−2	28.4	160	320.0
−1	30.2	170	338.0
0	32.0	180	356.0
1	33.8	190	374.0
2	35.6	200	392.0
3	37.4	210	410.0

°C	°F	°C	°F
4	39.2	220	428.0
5	41.0	230	446.0
6	42.8	240	464.0
7	44.6	250	482.0
8	46.4	260	500.0
9	48.2	270	518.0

A4.2.2 Fahrenheit to Celsius

°F	°C	°F	°C	°F	°C
−60	−51.1	10	−12.2	260	126.7
−50	−45.6	20	−6.7	270	132.2
−40	−40.0	30	−1.1	280	137.8
−30	−34.4	40	4.4	290	143.3
−20	−28.9	50	10.0	300	148.9
−10	−23.3	60	15.6	310	154.4
−9	−22.8	70	21.1	320	160.0
−8	−22.2	80	26.7	330	165.6
−7	−21.7	90	32.2	340	171.1
−6	−21.1	100	37.8	350	176.7
−5	−20.6	110	43.3	360	182.2
−4	−20.0	120	48.9	370	187.8
−3	−19.4	130	54.4	380	193.3
−2	−18.9	140	60.0	390	198.9
−1	−18.3	150	65.6	400	204.4
0	−17.8	160	71.1	410	210.0
1	−17.2	170	76.7	420	215.6
2	−16.7	180	82.2	430	221.1
3	−16.1	190	87.8	440	226.7
4	−15.6	200	93.3	450	232.2
5	−15.0	210	98.9	460	237.8
6	−14.4	220	104.4	470	243.3
7	−13.9	230	110.0	480	248.9
8	−13.3	240	115.6	490	254.4
9	−12.8	250	121.1	500	260.0

A4.3   Other Conversion Tables

A4.3.1   Metric Weight Measures

10 milligram (mg)	= 1 centigram
10 centigram (cg)	= 1 decigram
10 decigram (dg)	= 1 gram
10 grams (g)	= 1 dekagram
10 dekagrams (dkg)	= 1 hectogram
10 hectograms (hg)	= 1 kilogram
10 kilograms (kg)	= 1 myriagram
10 myriagrams (myg)	= 1 quintal

A4.3.2   Pounds to Kilograms

Pounds	Kilograms
1	0.454
2	0.907
3	1.361
4	1.814
5	2.268
6	2.722
7	3.175
8	3.629
9	4.082
10	4.536
25	11.340
50	22.680
75	34.020
100	45.360

A4.3.3 Ounces to Kilograms

Ounces	Kilograms
1	0.028
2	0.057
3	0.085
4	0.113
5	0.142
6	0.170
7	0.198
8	0.227
9	0.255
10	0.283
11	0.312
12	0.340
13	0.369
14	0.397
15	0.425
16	0.454

A4.3.4 Inches to Millimeters

Inches	Millimeters	Inches	Millimeters
1	25.4	11	279.4
2	50.8	12	304.8
3	76.2	13	330.2
4	101.6	14	355.6
5	127.0	15	381.0
6	152.4	16	406.4
7	177.8	17	431.8
8	203.2	18	457.2
9	228.6	19	482.6
10	254.0	20	508.0

A4.3.5   Fractions to Decimals

Fractions	Decimal Equivalent	Fractions	Decimal Equivalent
1/16	0.0625	9/16	0.5625
1/8	0.1250	5/8	0.6250
3/16	0.1875	11/16	0.6875
1/4	0.2500	3/4	0.7500
5/16	0.3125	13/16	0.8125
3/8	0.3750	7/8	0.8750
7/16	0.4375	15/16	0.9375
1/2	0.5000	1	1.0000

Bibliography

1. W. Hofmann, *Vulcanization and Vulcanizing Agents*, Maclaren & Sons, London, 1967.
2. *Transactions and Proceedings of the Institution of the Rubber Industry*, UK, June 1963.
3. C.C. Davis and J.T. Blake, *The Chemistry and Technology of Rubber*, Reinhold, New York, 1937.
4. *Polysar Rubber Technology Guide*, 1976.
5. V.C. Chandrasekaran, "Air borne Rubber Seals," *Polymer Review*, 1983.
6. Brian J. Wilson, *British Compounding Ingredients for Rubber*, W. Heffer & Sons, Cambridge, UK, 1958.
7. Rubber World Magazine's *Blue Book*.
8. *Bayer Products for the Rubber Industry*, Handbook.
9. V.C. Chandrasekaran, "Protection Against Corrosion—Elastomers," published articles.
10. Lawrence A. Wood, *Physical Constants of Rubbers*, National Institute of Science and Technology, Washington, DC.

Index

179

Plastics Design Library

Founding Editor: William A. Woishnis

The Effects of UV Light and Weather on Plastics and Elastomers, 2nd Ed., L. K. Massey, 978-0-8155-1525-8, 488 pp., 2007

Fluorinated Coatings and Finishes Handbook: The Definitive User's Guide and Databook, Laurence W. McKeen, 978-0-8155-1522-7, 400 pp., 2006

Fluoroelastomers Handbook: The Definitive User's Guide and Databook, Albert L. Moore, 0-8155-1517-0, 359 pp., 2006

Reactive Polymers Fundamentals and Applications: A Concise Guide to Industrial Polymers, J.K. Fink, 0-8155-1515-4, 800 pp., 2005

Fluoropolymers Applications in Chemical Processing Industries, P. R. Khaladkar, and S. Ebnesajjad, 0-8155-1502-2, 592 pp., 2005

The Effect of Sterilization Methods on Plastics and Elastomers, 2nd Ed., L. K. Massey, 0-8155-1505-7, 408 pp., 2005

Extrusion: The Definitive Processing Guide and Handbook, H. F. Giles, Jr., J. R. Wagner, Jr., and E. M. Mount, III, 0-8155-1473-5, 572 pp. 2005

Film Properties of Plastics and Elastomers, 2nd Ed., L. K. Massey, 1-884207-94-4, 250 pp. 2004

Handbook of Molded Part Shrinkage and Warpage, J. Fischer, 1-884207-72-3, 244 pp., 2003

Fluoroplastics, Volume 2: Melt-Processible Fluoroplastics, S. Ebnesajjad, 1-884207-96-0, 448 pp., 2002

Permeability Properties of Plastics and Elastomers, 2nd Ed. L. K. Massey, 1-884207-97-9, 550 pp., 2002

Rotational Molding Technology, R. J. Crawford and J. L. Throne, 1-884207-85-5, 450 pp., 2002

Specialized Molding Techniques & Application, Design, Materials and Processing, H. P. Heim, and H. Potente, 1-884207-91-X, 350 pp., 2002

Chemical Resistance CD-ROM, 3rd Ed., Plastics Design Library Staff, 1-884207-90-1, 2001

Plastics Failure Analysis and Prevention, J. Moalli, 1-884207-92-8, 400 pp., 2001

Fluoroplastics, Volume 1: Non-Melt Processible Fluoroplastics, S. Ebnesajjad, 1-884207-84-7, 365 pp., 2000

Coloring Technology for Plastics, R. M. Harris, 1-884207-78-2, 333 pp., 1999

Conductive Polymers and Plastics in Industrial Applications, L. M. Rupprecht, 1-884207-77-4, 302 pp., 1999

Imaging and Image Analysis Applications for Plastics, B. Pourdeyhimi, 1-884207-81-2, 398 pp., 1999

Metallocene Technology in Commercial Applications, G. M. Benedikt, 1-884207-76-6, 325 pp., 1999

Weathering of Plastics, G. Wypych, 1-884207-75-8, 325 pp., 1999

Dynamic Mechanical Analysis for Plastics Engineering, M. Sepe, 1-884207-64-2, 230 pp., 1998

Medical Plastics: Degradation Resistance and Failure Analysis, R. C. Portnoy, 1-884207-60-X, 215 pp., 1998

Metallocene Catalyzed Polymers, G. M. Benedikt and B. L. Goodall, 1-884207-59-6, 400 pp., 1998

Polypropylene: The Definitive User's Guide and Databook, C. Maier and T. Calafut, 1-884207-58-8, 425 pp., 1998

Handbook of Plastics Joining, Plastics Design Library Staff, 1-884207-17-0, 600 pp., 1997

Fatigue and Tribological Properties of Plastics and Elastomers, Plastics Design Library Staff, 1-884207-15-4, 595 pp., 1995

Chemical Resistance, Vol. 1, Plastics Design Library Staff, 1-884207-12-X, 1100 pp., 1994

Chemical Resistance, Vol. 2, Plastics Design Library Staff, 1-884207-13-8, 977 pp., 1994

The Effect of Creep and Other Time Related Factors on Plastics and Elastomers, Plastics Design Library Staff, 1-884207-03-0, 528 pp., 1991

The Effect of Temperature and Other Factors on Plastics, Plastics Design Library Staff, 1-884207-06-5, 420 pp., 1991

Printed and bound by CPI Group (UK) Ltd, Croydon, CR0 4YY

08/05/2025

01864832-0002